普通高等教育"十三五"规划教材

材料科学导论（双语）

Introduction to Materials Science(Bilingualism)

主　编　傅小明　蒋　萍
副主编　杨在志　李金涛　孙　虎

拓展学习资源

南京大学出版社

图书在版编目(CIP)数据

材料科学导论：双语 / 傅小明，蒋萍主编. —南京：南京大学出版社，2018.2
 ISBN 978-7-305-19362-0

Ⅰ. ①材… Ⅱ. ①傅… ②蒋… Ⅲ. ①材料科学-教材 Ⅳ. ①TB3

中国版本图书馆 CIP 数据核字(2017)第 246243 号

出版发行	南京大学出版社
社　　址	南京市汉口路 22 号　　邮　编 210093
出 版 人	金鑫荣
书　　名	材料科学导论（双语）
主　　编	傅小明　蒋　萍
副 主 编	杨在志　李金涛　孙　虎
责任编辑	秦　露　王日俊
照　　排	南京南琳图文制作有限公司
印　　刷	常州市武进第三印刷有限公司
开　　本	787×1092　1/16　印张 15.75　字数 417 千
版　　次	2018 年 2 月第 1 版　2018 年 2 月第 1 次印刷
ISBN	978-7-305-19362-0
定　　价	48.00 元

网址：http://www.njupco.com
官方微博：http://weibo.com/njupco
官方微信号：njupress
销售咨询热线：(025) 83594756

＊ 版权所有，侵权必究
＊ 凡购买南大版图书，如有印装质量问题，请与所购
　图书销售部门联系调换

前　言

随着人类文明的进步和科学技术的发展,材料已成为国民经济的三大支柱产业之一。为了适应材料科学与技术的发展,培养学生及时跟踪国际材料科学与技术发展前沿的能力,使学生成为材料科学领域的创新型复合人才,因而国内许多高校面向材料类本科生开设了双语专业课程。

目前,材料类双语专业课程《材料概论(双语)》或《材料科学与工程导论(双语)》教材尽管也有出版,但存在如下的主要问题:(1) 系统性不强,即知识点不全,不利于学生系统和全面的掌握相关知识;(2) 部分教材有少许课后练习题,但是没有参考答案,这样不利于学生课后对学习效果的检测和评价;(3) 更有甚者是半中文半英文,这样更不利于学生学习利用纯正的英语表述专业知识。这样难以满足培养应用型人才教学的要求。

编者基于上述三个问题得到如下基本认识:一是现代材料类高素质创新型人才标准——国际化、工程化和复合化的要求;二是基础应用性材料科学知识体系:科学性、系统性和简易性的特色;三是各类不同层次高校学生的教学要求。本教材针对低年级材料类大学生的实际条件和需求,在满足国际化专业人才语言交流能力的基本前提下,特别强调了专业基础知识体系的科学化、系统化和简易化,从而新编了《材料科学导论(双语)》和《材料工程导论(双语)》系列教材,以满足新时代各类特别是技术应用型材料科学与工程学科专业教学的需要。

本套教材具有如下特点:

1. 将材料学知识英语化

用英语语言思维构筑学生的国际化视野和专业语言交流能力。

2. 对知识性体系简易化

在保持教学内容的科学性、系统性的前提下,做到学生理解和掌握的简易性(通俗性),即突出"真、全、简"三个字。

3. 优化教学流程和效果

(1) 任务驱动教学

每章主要包括主要知识点的介绍、专业词汇的注解和课后练习题(附参考答案),充分体现了教学内容的实用性,有助于提高学生牢固掌握本章知识点的实践能力。

(2) 教材定位准确

本套教材针对材料类学生学习专业基础课后开设的课程,有助于学生应用英语去

前 言

描述自己本专业的材料学知识。

（3）内容结构合理

本套教材内容由浅入深，循序渐进，符合读者认识事物的规律性。同时，也便于教学的组织、实施和考核，有利于教学效果的巩固和教学质量的提高。

《材料科学导论（双语）》教材是由宿迁学院材料工程系傅小明副教授、江苏大学外国语学院蒋萍老师担任主编，并编写绪论（第1章）、第一篇材料相变基础的第2、3和4章、第三篇材料性能基础的第8章，以及全书的统稿；杨在志老师编写第二篇材料结构基础的第5章和第三篇材料性能基础的第9章；李金涛老师编写第二篇材料结构基础的第6和7章；孙虎老师编写第三篇材料性能基础的第10章。江苏大学外国语学院蒋萍老师编写全书专业术语和专业词汇的注解，以及全书习题及其参考答案。

本书在编写过程中得到了兰州理工大学博士生导师马勤教授悉心的指导，在此特表感谢。

由于编者水平有限，经验不足，书中难免有不足之处，恳请专家、学者和广大读者批评指正。

编 者

2017 年 04 月

Contents

Chapter 1　Introduction ································· 1
  1.1　Historical Perspective ································· 1
  1.2　Materials ································· 2
  1.3　Classification of Materials ································· 2
    1.3.1　Metals ································· 3
    1.3.2　Ceramics ································· 3
    1.3.3　Polymers ································· 3
    1.3.4　Composites ································· 3
    1.3.5　Advanced Materials ································· 4
      1.3.5.1　Semiconductors ································· 4
      1.3.5.2　Biomaterials ································· 4
      1.3.5.3　Smart Materials ································· 4
      1.3.5.4　Nanoengineered Materials ································· 5
  1.4　Structural Characteristic of Materials ································· 6
    1.4.1　Crystal Lattice ································· 7
    1.4.2　Crystallographic Indices ································· 11
    1.4.3　Anisotropy ································· 12
  1.5　Materials Properties ································· 12
    1.5.1　Physical and Chemical Properties of Materials ································· 12
    1.5.2　Mechanical Properties of Materials ································· 15
  1.6　Materials Science ································· 18
  1.7　Modern Materials' Needs ································· 20

Part I　Foundation of Phase Change of Materials

Chapter 2　Phase Diagrams ································· 29
  2.1　Introduction ································· 29
  2.2　Definitions and Basic Concepts ································· 29
  2.3　Solubility Limit ································· 30
  2.4　Phases ································· 31

2.5 Phase Equilibrium ········· 32
2.6 Equilibrium Phase Diagrams ········· 34
2.7 Interpretation of Phase Diagrams ········· 36
    2.7.1 Phases Present ········· 36
    2.7.2 Determination of Phase Compositions ········· 36
    2.7.3 Determination of Phase Amounts ········· 37
    2.7.4 Binary Eutectic Systems ········· 39
    2.7.5 Gibbs Phase Rule ········· 41
    2.7.6 Iron-Iron Carbide Phase Diagram ········· 44

Chapter 3 Solidification and Crystallization ········· 51

3.1 Introduction ········· 51
3.2 Solidification of Metals ········· 51
    3.2.1 Formation of Stable Nuclei in Liquid Metals ········· 52
    3.2.2 Homogeneous Nucleation ········· 52
    3.2.3 Critical Radius and Undercooling ········· 54
    3.2.4 Heterogeneous Nucleation ········· 55
3.3 Growth of Crystals ········· 56
    3.3.1 Growth of Crystals in Liquid Metal and Formation of a Grain Structure
    ········· 56
    3.3.2 Solidification of Single Crystals ········· 57
    3.3.3 Metallic Solid Solutions ········· 58
        3.3.3.1 Substitutional Solid Solutions ········· 59
        3.3.3.2 Interstitial Solid Solutions ········· 59

Chapter 4 Phase Transformation ········· 63

4.1 Introduction ········· 63
4.2 Phase Transformation ········· 64
    4.2.1 Kinetics of Solid-State Reaction ········· 64
    4.2.2 Multiphase Transformations ········· 66
4.3 Microstructural and Property Changes in Iron-Carbon Alloys ········· 67
    4.3.1 Isothermal Transformation Diagrams ········· 67
    4.3.2 Continuous Cooling Transformation Diagram ········· 75

Part II Foundation of Material Structures

Chapter 5 Crystal Structure ········· 85

5.1 Introduction ········· 85
5.2 Fundamental Concepts ········· 85

    5.2.1 Space Lattice and Unit Cells ······ 85
    5.2.2 Crystal Systems and Bravais Lattice ······ 86
    5.2.3 Crystallographic Directions and Miller Indies ······ 88
        5.2.3.1 Atom Positions in Unit Cells ······ 88
        5.2.3.2 Directions in Cubic Unit Cells ······ 89
        5.2.3.3 Miller Indices for Crystallographic Planes in Cubic Unit Cells ······ 90
        5.2.3.4 Crystallographic Planes and Directions in Hexagonal Unit Cells ······ 93
5.3 Principal Metallic Crystal Structures ······ 95
    5.3.1 Body-Centered Cubic (BCC) Crystal Structure ······ 96
    5.3.2 Face-Centered Cubic (FCC) Crystal Structure ······ 98
    5.3.3 Hexagonal Close-Packed (HCP) Crystal Structure ······ 99
    5.3.4 Comparison of FCC, HCP and BCC Crystal Structures ······ 101
        5.3.4.1 FCC and HCP ······ 101
        5.3.4.2 BCC ······ 102
    5.3.5 Volume, Planar and Linear Density Unit-Cell Calculations ······ 103
        5.3.5.1 Volume Density ······ 103
        5.3.5.2 Planar Atomic Density ······ 103
        5.3.5.3 Linear Atomic Density ······ 104
    5.3.6 Polymorphism or Allotropy ······ 104

Chapter 6 Defect Structure ······ 109

6.1 Introduction ······ 109
6.2 Point Defects ······ 109
    6.2.1 Point Defects in Metals ······ 109
    6.2.2 Point Defects in Ceramics ······ 110
    6.2.3 Impurities in Solids ······ 112
        6.2.3.1 Impurities in Metals ······ 112
        6.2.3.2 Solid Solutions ······ 113
        6.2.3.3 Impurities in Ceramics ······ 115
    6.2.4 Point Defects in Polymers ······ 116
    6.2.5 Specification of Composition ······ 116
6.3 Miscellaneous Imperfections ······ 117
    6.3.1 Dislocations-Linear Defects ······ 117
    6.3.2 Interfacial Defects ······ 120
        6.3.2.1 External Surfaces ······ 120
        6.3.2.2 Grain Boundaries ······ 120
        6.3.2.3 Twin Boundaries ······ 121
        6.3.2.4 Miscellaneous Interfacial Defects ······ 122

6.3.3 Bulk or Volume Defects ……… 122
  6.3.4 Atomic Vibrations ……… 123

Chapter 7 Structure of Bulk Phase ……… 126
 7.1 Introduction ……… 126
 7.2 Single Crystals ……… 126
 7.3 Polycrystalline Materials ……… 127
 7.4 Noncrystalline Solids ……… 128
 7.5 Quasicrystals ……… 129

Part III Foundation of Material Properties

Chapter 8 Mechanical Properties of Materials ……… 137
 8.1 Introduction ……… 137
 8.2 Concepts of Stress and Strain ……… 138
  8.2.1 Tension Tests ……… 138
  8.2.2 Compression Tests ……… 141
  8.2.3 Shear and Torsional Tests ……… 141
  8.2.4 Geometric Considerations of the Stress State ……… 142
 8.3 Elastic Deformation ……… 143
  8.3.1 Stress-Strain Behavior ……… 143
  8.3.2 Anelasticity ……… 147
  8.3.3 Elastic Properties of Materials ……… 147
 8.4 Mechanical Behavior of Metals ……… 149
  8.4.1 Tensile Properties ……… 149
   8.4.1.1 Yielding and Yield Strength ……… 149
   8.4.1.2 Tensile Strength ……… 151
   8.4.1.3 Ductility ……… 151
   8.4.1.4 Resilience ……… 154
   8.4.1.5 Toughness ……… 155
  8.4.2 True Stress and Strain ……… 156
  8.4.3 Elastic Recovery during Plastic Deformation ……… 158
  8.4.4 Compressive, Shear and Torsional Deformation ……… 159
 8.5 Mechanical Behavior-Ceramics ……… 159
  8.5.1 Flexural Strength ……… 159
  8.5.2 Elastic Behavior ……… 161
 8.6 Mechanical Behavior of Polymers ……… 161
  8.6.1 Stress-Strain Behavior ……… 161

  8.6.2 Macroscopic Deformation ······ 164
8.7 Hardness and Other Mechanical Property Considerations ······ 165
  8.7.1 Hardness ······ 165
  8.7.2 Rockwell Hardness Tests ······ 166
  8.7.3 Brinell Hardness Tests ······ 169
  8.7.4 Knoop and Vickers Microhardness Tests ······ 170
  8.7.5 Hardness Conversion ······ 170
  8.7.6 Correlation between Hardness and Tensile Strength ······ 171
  8.7.7 Hardness of Ceramic Materials ······ 171
  8.7.8 Tear Strength and Hardness of Polymers ······ 171
8.8 Property Variability and Design/Safety Factors ······ 172
  8.8.1 Variability of Material Properties ······ 172
  8.8.2 Design/Safety Factors ······ 173

Chapter 9 Physical Properties of Materials ······ 177

9.1 Introduction ······ 177
9.2 Electrical Properties of Materials ······ 178
  9.2.1 Metals and Alloys ······ 179
  9.2.2 Semiconductors ······ 180
    9.2.2.1 Intrinsic Semiconductors ······ 181
    9.2.2.2 Extrinsic Semiconductors ······ 182
    9.2.2.3 Compound Semiconductors ······ 183
  9.2.3 Ionic Ceramics and Polymers ······ 183
9.3 Thermal Properties of Materials ······ 184
  9.3.1 Heat Capacity ······ 184
  9.3.2 Thermal Expansion ······ 185
    9.3.2.1 Metals ······ 185
    9.3.2.2 Ceramics ······ 186
    9.3.2.3 Polymers ······ 186
  9.3.3 Thermal Conductivity ······ 186
  9.3.4 Thermal Stresses ······ 187
    9.3.4.1 Stresses Resulting from Restrained Thermal Expansion and Contraction ······ 187
    9.3.4.2 Stresses Resulting from Temperature Gradients ······ 187
    9.3.4.3 Thermal Shock of Brittle Materials ······ 188
9.4 Magnetic Properties of Materials ······ 189
  9.4.1 Diamagnetism, Paramagnetism and Ferromagnetism ······ 189
  9.4.2 Antiferromagnetism and Ferrimagnetism ······ 192

Contents

    9.4.2.1 Antiferromagnetism ································ 192
    9.4.2.2 Ferrimagnetism ··································· 193
  9.4.3 The Influence of Temperature on Magnetic Behavior ············ 195
  9.4.4 Domains, Hysteresis and Magnetic Anisotropy ················ 196
  9.4.5 Superconductivity ····································· 199
 9.5 Optical Properties of Materials ································ 204
  9.5.1 Interaction of Light with Matter ·························· 205
  9.5.2 Atomic and Electronic Interactions ······················· 206
    9.5.2.1 Electronic Polarization ·························· 206
    9.5.2.2 Electron Transitions ···························· 206
    9.5.2.3 Optical Properties of Metals ····················· 208
    9.5.2.4 Optical Properties of Nonmetals ··················· 209
  9.5.3 Refraction, Reflection, Absorption and Transmission ········· 209
  9.5.4 Opacity and Translucency in Insulators ···················· 211
  9.5.5 Applications of Optical Phenomena ······················ 212
    9.5.5.1 Luminescence ································ 212
    9.5.5.2 Photoconductivity ····························· 212

Chapter 10 Chemical Properties of Materials ············· 217

 10.1 Introduction ·· 217
 10.2 Corrosion of Metals ··· 217
  10.2.1 Cost of Corrosion in Industry ··························· 218
  10.2.2 Classification of Corrosion ····························· 218
  10.2.3 Corrosion Mechanism ································· 219
  10.2.4 Electrochemical Considerations ························· 222
  10.2.5 Corrosion Rates ····································· 225
  10.2.6 Passivity ··· 225
  10.2.7 Environmental Effects ································ 225
  10.2.8 Forms of Corrosion ·································· 226
  10.2.9 Corrosion Environments ······························ 232
  10.2.10 Corrosion Prevention ······························· 232
 10.3 Corrosion of Ceramic Materials ······························· 233
 10.4 Degradation of Polymers ····································· 234
  10.4.1 Swelling and Dissolution ······························ 234
  10.4.2 Bond Rupture ······································· 235
  10.4.3 Weathering ··· 236

Main References ·· 240

Chapter 1 Introduction

1.1 Historical Perspective

Materials are probably more deep-seated① in our culture than most of us realize. Transportation, housing, clothing, communication, recreation and food production-virtually every segment of our everyday lives is influenced to one degree or another by materials. Historically, the development and advancement of societies have been intimately tied to the members' ability to produce and manipulate materials to fill their needs. In fact, early civilizations have been designated by the level of their materials development (*i.e.*, Stone Age, Bronze Age).

The earliest humans had access to only a very limited number of materials, those than occur naturally: stone, wood, lay, skins, and so on. With time they discovered techniques for producing materials that had properties superior to those of the natural ones; these new materials included pottery and various metals. Furthermore, it was discovered that the properties of a materials could be altered by heat treatment and by the addition of other substances. At the point, materials utilization was totally a selection process, that is, deciding from a given, rather limited set of materials the one that was best suited for an application by virtue of its characteristics. It was not until relatively recent times that scientists came to understand the relationships between the structural elements of materials and their properties. This knowledge acquired in the past 60 years or so, has empowered them to fashion, to a large degree, the characteristics of materials. Thus, tens of thousands of different materials have evolved with rather specialized characteristics that meet the needs of our modern and complex society; these include metals, plastics, glasses and fibers.

The development of many technologies that make our existence so comfortable has been intimately associated with the accessibility of suitable materials. An advancement in the understanding of a material type is often the forerunner to the

Chapter 1 Introduction

stepwise progression of a technology. For example, automobiles would not have been possible without the availability of inexpensive steel or some other comparable substitute. In our contemporary era, sophisticated electronic devices rely on components that are made from what are called semiconducting materials.

1.2 Materials

The materials making up the surrounding world consist of discrete particles, having a submicroscopic size. Their behavior is determined by atomic theories. The states of organization of materials range from the complete disorder of atoms or molecules of a gas under weak pressure to the almost complete order of atoms in a monocrystal.

In this introductory work materials are defined as solids used by man to produce items which constitute the support for his living environment.

Indeed, no object exists without materials. All sectors of human activity depend on materials, from the manufacture of an integrated circuit to the construction of a hydroelectric dam. They appear in our bodies to strengthen or replace our damaged biomaterials. Materials are also as indispensable to our society as food, energy and information. Their essential role is too often forgotten.

The definition employed in this introductory work is limited to solid materials. It excludes liquids and gases, as well as solid combustibles.

1.3 Classification of Materials

Solid materials have been conveniently grouped into three basic classifications: metals, ceramics and polymers. This scheme is based primarily on chemical makeup and atomic structure, and most materials fall into one distinct grouping or another, although there are some intermediates. What may be considered to be a fourth group—the composites-consists of combinations of two or more different materials. Another classification is advanced materials-those used in high-technology applications (*viz*, semiconductors, biomaterials, smart materials and nanoengineered materials). A brief explanation of the material types and representative characteristics is offered next.

1.3.1 Metals

Metallic materials are normally combinations of metallic elements. They have large numbers of nonlocalized electrons, that is, these electrons are not bound to particular atoms. Many properties of metals are directly attributable to these electrons. Metals are extremely good conductors of electricity and are not transparent to visible light; a polished metal surface has a lustrous appearance. Furthermore, metals arquite strong, yet deformable, which accounts for their extensive use in structural applications.

1.3.2 Ceramics

Ceramics are compounds between metallic and nonmetallic elements; they are most frequently oxides, nitrides and carbides. The wide range of materials that falls within this classification includes ceramics that are composed of clay minerals, cement and glass. These materials are typically insulative to the passage of electricity and heat, and are more resistant to high temperatures and harsh environments than metals and polymers. With regard to mechanical behavior, ceramics are hard but very brittle.

1.3.3 Polymers

Polymers include the familiar plastic and rubber materials. Many of them are organic compounds that are chemically based on carbon, hydrogen and other nonmetallic elements; furthermore, they have very large molecular structures. These materials typically have low densities and may be extremely flexible.

1.3.4 Composites

A number of composite materials have been engineered that consist of more than one material type. Fiberglass is a familiar example, in which glass fibers are embedded within a polymeric material. A composite id designed to display a combination of the best characteristics of each of the component material. Fiberglass acquires strength from the glass and flexibility from the polymer. Many of the recent material developments have involved composite materials.

1.3.5 Advanced Materials

Materials that are utilized in high-technology (or high-tech) applications are sometimes termed advanced materials. By high technology we mean a device or product that operates or functions using relatively intricate and sophisticated principles; examples include electronic equipment (camcorders, CD/DVD players, etc.), computers, fiber-optic systems, spacecraft, aircraft and military rocketry. These advanced materials are typically traditional materials whose properties have been enhanced, and also newly developed, high-performance materials. They may be of all material types (e.g., metals, ceramics, polymers), and are normally expensive. Advanced materials include semiconductors, biomaterials, and what we may term "materials of the future" (that is, smart materials and nanoengineered materials).

1.3.5.1 Semiconductors

Semiconductors have electrical properties that are intermediate between the electrical conductors and insulators. Furthermore, the electrical characteristics of these materials are extremely sensitive to the presence of minute concentrations of impurity atoms, which concentrations may be controlled over very small spatial regions. The semiconductors have made possible the advent of integrated circuitry that has totally revolutionized the electronics and computer industries (not to mention our lives) over the past two decades.

1.3.5.2 Biomaterials

Biomaterials are employed in components implanted into the human body for replacement of diseased or damaged body parts. These materials must not produce toxic substances and must be compatible with body tissues (i.e., must not cause adverse biological reactions). All of the above materials-metals, ceramics, polymers, composites and semiconductors may be used as biomaterials. For example, some of the biomaterials that are utilized in artificial hip replacements.

1.3.5.3 Smart Materials

Smart (or intelligent) materials are a group of new and state-of-the-art materials now being developed that will have a significant influence on many of our technology. The adjective "smart" implies that these materials are able to sense changes in their environments and then respond to these changes in predetermined manners-traits that are also found in living organisms. In addition, this "smart" concept is being extended to rather sophisticated systems that consist of both smart and traditional materials.

Components of a smart material (or system) include some type of sensor (that detects an input signal), and an actuator (that performs a responsive and adaptive function). Actuators may be called upon to change shape, position, natural frequency or mechanical characteristics in response to changes in temperature, electric fields, and/or magnetic fields. Four types of materials are commonly used for actuators: shape memory alloys, piezoelectric ceramics, magnetostrictive materials and electroheological/magnetorheological fluids. Shape memory alloys are metals that, after having been deformed, revert back to their original shapes when temperature is changed. Piezoelectric ceramics expand and contract in response to an applied electric field (or voltage); conversely, they also generate an electric field when their dimensions are altered. The behavior of magnetostrictive materials is analogous to that of the piezoelectrics, except that they are responsive to magnetic fields. Also, electrorheological and magnetorheological fluids are that experience dramatic changes in viscosity upon the application of electric and magnetic fields, respectively.

Materials/devices employed as sensors include optical fibers, piezoelectricmaterials (including some polymers) and microelectromechanical devices.

For example, one type of smart system is used in helicopters to reduce aerodynamic cockpit noise that is created by the rotating rotor blades. Piezoelectric sensors inserted into the blades monitor blade stresses and deformations; feedback signals from these sensors are fed into a computer-controlled adaptive device, which generates noise-canceling[③] antinoise.

1.3.5.4 Nanoengineered Materials

Until very recent times the general procedure utilized by scientists to understand the chemistry and physics of materials has been to begin by studying large and complex structures, and then to investigate the fundamental building blocks of these structures that are smaller and simpler. Thisapproach is sometimes termed "top-down[④]" science. However, with the advent of scanning probe microscopes, which permit observation of individual atoms and molecules, it has become possible to manipulate and move atoms and molecules to form new structures, thus, design new materials that are built from simple atomic-level constituents (i.e., "materials by design"). This ability to carefully arrange atoms provides opportunities to develop mechanical, electrical, magnetic and other properties that are not otherwise possible. We call this the "bottom-up[5]" approach, and the study of the properties of these materials is termed "nanotechnology". The "nano" prefix denotes that the dimensions of these structural entities are on the order of a nanometer (10^{-9} m) as a rule, less than 100 nanometers (equivalent to approximately 500 atom diameters). One example of a material of this type is the carbon nanotube. In the future we will undoubtedly find that increasingly more of our technological advances will utilize these nanoengineered materials.

1.4 Structural Characteristic of Materials

Solids exist in nature in two principal forms: crystalline and amorphous, which differ substantially in their properties.

Crystalline bodies remain solid, i.e. retain their shape, up to a definite temperature (melting point) at which they change from the solid to liquid state (Figure 1.1). During cooling, the inverse process of solidification takes place, again at the definite solidifying temperature, or point. In both cases, the temperature remains constant until the material is completely melted or respectively solidified.

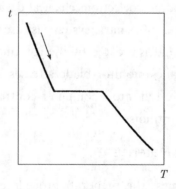

Figure 1.1　Cooling curve of a crystalline substance

Amorphous bodies, which heated, are graduallysoftened in a wide temperature range and become viscous and only then change to the liquid state. In cooling, the process takes place in the opposite direction.

The crystalline state of a solid is more stable than amorphous state.

Amorphous bodies differ from liquid in having a lower mobility of particles. An amorphous state can be fixed in many organic and inorganic substances by rapid cooling from the liquid state. On repeated heating long holding at 20~25 ℃ or, in some cases, deformation of an amorphous body, the instability of the amorphous state may result in a partial or complete change to the crystalline state.

Examples of such changes from amorphous to crystalline state are turbidity effect appearing in inorganic glasses on heating or in optical glasses after a long use, partial crystallization of molten amber on heating, or additional crystallization and strengthening of nylon fibres on tension.

Crystalline bodies are characterized by an ordered arrangement of their elementary particles (ions, atoms or molecules). The properties of crystals depend on the electronic structure of atoms and the nature of their interactions in the crystal, on the spatialarrangement of elementary particles, and on the composition, size and shape of crystals.

The structure of crystals is described by using the concepts of fine structure and micro- and macro-structure depending on the size of structural components and methods employed to reveal them.

Microscopic examinations make it possible to determine the size and shape of grains (crystals), the presence of different nature, their distribution and relative volume quantities, the shape of foreign inclusions and microvoids, orientations of crystals, and some special crystallographic characteristics (twins, slip lines, *etc.*).

Macrostructure of crystals is studied by the naked eye or with a magnifying glass. This method can reveal the pattern of a fracture, shrinkage cavities and voids, and the shape and size of large crystals. It is also possible to detect cracks, chemical inhomogeneities, fibrous textures, *etc.* by using specially prepared (polished and etched) specimens. Macrostructural examination is a valuable method for studying crystalline materials.

1.4.1　Crystal Lattice

In a model of a crystal, the elementary particles (ions, atoms or molecules) that constitute its structure can be imagined to be spheres which touch one another and are arranged regularly in different directions (Figure 1.2 (a)). In a simpler model of crystal structure, spheres are replaced by points representing the centres of particles (Figure 1.2 (b)).

If three directions, x, y and z, belonging to different planes, are drawn in a crystal, the spacings between the particles arranged along these directions will in the general case be different (say, a, b and c).

Planes parallel to the coordinate planes and spaced at distances a, b and c from one another and oriented in parallel. The smallest of such parallelepipeds is called an elementary cell. Successive displacements of such a parallelepiped can form a

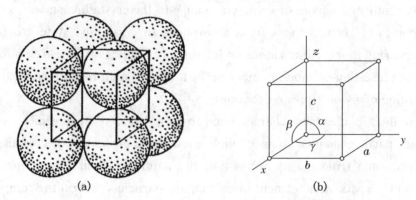

Figure 1.2　Arrangment of elementary particles in a crystal

three-dimensional crystal lattice. The corners of the parallelepiped are called sites of a crystal lattice. These sites coincide with the centres of the particles which the crystal is built of.

A three-dimensional crystal lattice defines completely the structure of a crystal.

An elementary cell of the crystal lattice is described by three sections a, b and c, which are equal to the distances to the nearest elementary particles along the coordinate axes and three angles made by each two of these sections, α, β and γ.

The dimensions of an elementary cell of the crystal lattice are determined by sections a, b and c. They are called lattice spacings (or lattice constants). With known spacings of a lattice, it is possible to determine the ionic or atoms radius of an element. It is half the shortest spacing between particles in a lattice.

In most cases, crystal lattices have an intricate order, since elementary particles can occupy not only the lattice sites, but also be arranged on its faces or in the centre (Figure 1.3). The complexity of a lattice is decided by the number of particles per elementary cell. In a simple three-dimensional lattice (Figure 1.3(a)), a single particle always falls per cell. Each cell has eight corners, but each particle in a corner is shared simultaneously by eight other cells, roughly 1/8 of the volume of a site falls on each cell and, since there are eight sites in a cell, one elementary particle falls on an elementary cell. In complex three-dimensional lattices, the number of particles per cell is always more than one. In a body-centred cell (Figure 1.3(b)), there are two particles: one from a corner and one centring particle which belongs solely to this particular cell. In a face-centred cell (Figure 1.3(c)) there are four particles: one from the corners and three from six centred planes (since an elementary particle in the centre of a face plane is shared simultaneously by two cells).

A system, spacing and the number of particles per elementary cell determine uniquely the arrangement of elementary particles in a crystal.

In some cases, additional characteristics of crystal lattices and used, which follow from the lattice geometry and reflect the packing density of elementary particles in a crystal. Among such characteristics are the coordination number and the packing factor.

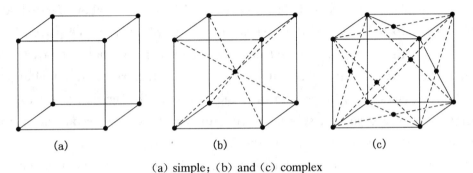

(a) simple; (b) and (c) complex

Figure 1.3 Types of elementary cell in crystal lattices

The coordination number determines the quantity of the nearest equidistant elementary particles. For instance, in a body-centred cubic (BCC) lattice, the number of such neighbors for each atom is eight (C8). In a simple cubic lattice, the coordination number is six (C6). In a face-centred cubic (FCC) lattice, the coordination number is twelve (C12).

The packing factor is determined as the ratio of the volume of all elementary particles per elementary cell to the total volume of the elementary cell. This factor is equal to 0.52 for a simple cubic lattice, 0.68 for a body-centred cubic lattice and 0.74 for a face-centred cubic lattice.

The remaining space of an elementary cell is occupied by interstitial voids or interstices which are differentiated into octahedral and tetrahedral.

The centers of such voids of an FCC lattice are shown by small dots in Figure 1.4. The radius of an octahedral void is 0.41 of the radius of the elementary particle and the radius of a tetrahedral void is equal to 0.22.

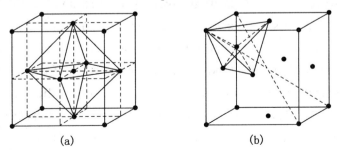

Figure 1.4 Octahedral (a) and tetrahedral (b) voids in FCC lattice metals

Chapter 1　Introduction

Close packing of elementary particles is typical of many crystals. If elementary particles are represented by spheres (which is true for most particle since they possess spherical symmetry), they can be packed into a number of structures as shown in Figure 1.5.

In that figure, the first layer is formed by spheres A which are close-packed[①] in a hexagonal plane. The second layer (spheres B) is placed above it so that its spheres get into recesses 1 of the first layer. The third layer can be placed in two ways: (1) if its spheres are placed directly above those of the first layer, there forms a hexagonal close-packed lattice which is characterized by an alternating *ABAB* arrangement of spheres (shown at the bottom of the figure); (2) if the third layer (C) is placed so that spheres C fit into recesses of the second layer above recesses 2 of the first layer and only the fourth layer repeats exactly the first one, there forms a face-centred cubic lattice with an *ABCABC* arrangement of spheres.

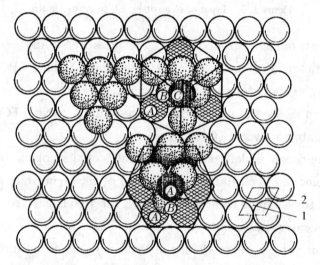

Figure 1.5　Close packing of atoms in crystal

In that figure, the first layer is formed by spheres A which are close-packed in a hexagonal plane. The second layer (spheres B) is placed above it so that its spheres get into recesses 1 of the first layer. The third layer can be placed in two ways: (1) if its spheres are placed directly above those of the first layer, there forms a hexagonal close-packed lattice which is characterized by an alternating *ABAB* arrangement of spheres (shown at the bottom of the figure); (2) if the third layer (C) is placed so that spheres C fit into recesses of the second layer above recesses 2 of the first layer and only the fourth layer repeats exactly the first one, there forms a face-centred cubic lattice with an *ABCABC* arrangement of spheres.

1.4.2 Crystallographic Indices

The properties of a crystal are the same along parallel directions. Therefore, it suffices to indicate a single direction passing through the origin of coordinates for a whole family of parallel lines. This makes it possible to define the direction of a line by a single point, since the other point is always the origin of coordinates. Such a reference point may be the site of a crystal lattice occupied by an elementary particle. The coordinates of this site are expressed by whole number u, v and w measured in the units of sections a, b, c, and written in square brackets: [u, v and w]; they are called direction indices. A negative index is designated by a bar above the number (Figure 1.6(a)).

The position of a plane in space is determined by sections cut off by that plane on the axes x, y and z. These sections are given by whole numbers m, n and p measured in the units of sections a, b and c. It has been adopted to use their reciprocals as plane indices: $h = 1/m$; $k = 1/n$ and $l = 1/p$. The numbers h, k and l written in parentheses, are called the plane indices, or Miller indices[1] Figure 1.6 (b). If negative sections are cut off by the plane, this is indicated by a bar above the corresponding index.

Close-packed planes are called slip planes, since they are the planes of preferable displacement of atoms in a crystal during plastic deformation.

For FCC crystals, the slip planes are those of the family (111). For HCP crystals with the c/a ratio equal to or more than 1.633, the slip plane is the basal plane, i.e. the hexagonal base of the prism. With $c/a < 1.633$, the planes of the prism are also the slip planes.

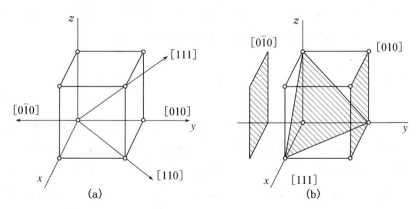

Figure 1.6 Crystallographic induces of directions (a) and planes (b)

Chapter 1 Introduction

1.4.3 Anisotropy

The properties of crystals are different in various crystallographic directions, which are associated with an ordered arrangement of atoms (ions and molecules) in space. The phenomenon is called anisotropy.

The properties of crystals are determined by interactions of atoms. In crystals, the spacings between atoms are different in various crystallographic directions, because of which their properties are also different.

Virtually all properties of crystals are anisotropic. The phenomenon is however more pronounced in crystals with structures of a poor symmetry.

Anisotropy of properties is mainly observed in single grown crystals. Natural crystalline solids are mostly polycrystals, i. e. they consist of a plurality of differently oriented fine crystals and exhibit no anisotropy, since the mean statistic spaings between atoms are essentially the same in all directions. In that connection, polycrystalline solids are considered to be quasi-isotropic. After plastic working of a polycrystal, crystallographic planes of the same index may turn out to be oriented in parallel. Such polycrystals are called textured, like single crystals, they are anisotropic.

1.5 Materials Properties

A material exhibits a set of properties, which define its behavior. A property of a material is determined by analyzing thereaction of the material to some outside influence, generally by means of a normalized standard test. According to the type of outside influence, two categories of properties are recognized.

1.5.1 Physical and Chemical Properties of Materials

Physical properties are those that can be observed without changing the identity of the substance. The general properties of matter such as color, density, hardness, are examples of physical properties. Properties that describe how a substance changes into a completely different substance are called chemical properties. Flammability and corrosion/oxidation resistance are examples of chemical properties.

The different between a physical and chemical property is straightforward until the phase of the materials is considered. When a material changes from a solid to a liquid to a vapor it seems like them become a difference substance. However, when a material melts, solidifies, vaporizes, condenses or sublimes, only the state of the substance changes. Considerice, liquid water and water vapor, they are all simply H_2O. Phase is a physical property of matter and matter can exist in four phases: solid, liquid, gas and plasma.

In general, some of the more important physical and chemical properties from an engineering material standpoint include phase transformation temperatures, density, specificgravity, thermal conductivity, linear coefficient of thermal expansion, electrical conductivity and resistivity, magnetic permeability and corrosion resistance, and so on.

(1) Phase transformation temperatures

When temperature rises and pressure is held constant, a typical substance changes from solid to liquid and then to vapor. Transitions from solid to liquid, from liquid to vapor, from vapor to solid and visa versa are called phase transformations or transitions. Since some substances have several crystal forms, technically there can also be solid to another solid form phase transformation.

Phase transitions from solid to liquid, and from liquid to vapor will absorb heat. The phase transition temperature where a solid changes to a liquid is called the melting point. The temperature at which the vapor pressure of a liquid equals 1 atm (101.3 kPa) is called the boiling point. Some materials, such as many polymers, do not go simply from a solid to a liquid with increasing temperature. Instead, at some temperature below the melting point, they start to lose their crystalline structure but the molecules remain linked in chains, which results in a soft and pliable material. The temperature at which a solid, glassy material begins to soften and flow is called the glass transition temperature.

(2) Density

Mass can be thinly distributed as in a pillow, or tightly packed as in a block of lead. The space the mass occupies is its volume, and the mass per unit of volume is its density.

Mass (m) is a fundamental measure of the amount of matter. Weight (ω) is a measure of the force exerted by a mass and this force is produced by the acceleration of gravity. Therefore, on the surface of the earth, the mass of an object is determined by dividing the weight of an object by 9.8 m/s² (the acceleration of gravity on the surface of the earth). Since we are typically comparing things on the surface of the earth, the weight of an object is commonly used rather than calculating its mass.

Chapter 1 Introduction

The density (r) of a material depends on the phase it is in and the temperature (The density of liquids and gases is very temperature dependent). Water in the liquid state has a density of 1 g/cm^3 = 1 000 g/m^3 at 40 ℃. Ice has a density of 0.917 g/cm^3 at 0 ℃, and it should be noted that this decrease in density for the solid phase is unusual. For almost all other substances, the density of the solid phase is greater than that of the liquid phase. Water vapor (vapor saturated air) has a density of 0.051 g/cm^3.

Some common units used for expressing density are grams/cubic centimeter, kilograms/cubic meter, grams/milliliter, grams/liter, pounds for cubic inch and pounds per cubic foot; but it should be obvious that any unit of mass per any unit of volume can be used.

(3) Specific gravity

Specific gravity is the ratio of density of a substance compared to the density of fresh water at 4 ℃ (40 ℉). At this temperature the density of water is at its greatest value and equal 1 g/cm^3. Since specific gravity is a ratio, so it has no units. An object will float in water if its density is less than the density of water and sink if its density is greater that that of water. Similarly, an object with specific gravity less than 1 will float and those with a specific gravity greater than one will sink. Specific gravity values for a few common substances are: Au, 19.3; mercury, 13.6; alcohol, 0.789 3; benzene, 0.878 6. Note that since water has a density of 1 g/cm^3, the specific gravity is the same as the density of the material measured in g/cm^3.

(4) Magnetic permeability

Magnetic permeability or simply permeability is the ease with which a material can be magnetized. It is a constant of proportionality that exists between magnetic induction and magnetic field intensity. This constant is equal to approximately 1.257 × 10^{-6} Henry per meter (H/m) in free space (a vacuum). In other materials it can be much different, often substantially greater than the free-space value, which is symbolized μ_0.

Materials that cause the lines of flux to move farther apart, resulting in a decrease in magnetic flux density compared with a vacuum, are called diamagnetic. Materials that concentrate magnetic flux by a factor of more than one but less than or equal to ten are called paramagnetic; materials that concentrate the flux by a factor of more than ten are called ferromagnetic. The permeability factors of some substances change with rising or falling temperature, or with the intensity of the applied magnetic field.

In engineering applications, permeability is often expressed in relative, rather than in absolute, terms. If μ_0 represents the permeability of free space (that is,

1.257×10⁻⁶ H/m) and μ represents the permeability of the substance in question (also specified in henrys per meter), then the relative permeability, μ_r, is given by through formula (1-1):

$$\mu_r = \mu/\mu_0 \tag{1-1}$$

For non-ferrous metals such as copper, brass, aluminum, *etc.*, the permeability is the same as that of "free space", *i.e.* the relative permeability is one. For ferrous metals however the value of μ_r may be several hundred. Certain ferromagnetic materials, especially powdered or laminated iron, steel, or nickel alloys, have μ_r that can range up to about 1 000 000. Diamagnetic materials have μ_r less than one, but no known substance has relative permeability much less than one. In addition, permeability can vary greatly within a metal part due to localized stresses, heating effects, *etc*.

When a paramagnetic or ferromagnetic core is inserted into a coil, the inductance is multiplied by μ_r compared with the inductance of the same coil with an air core. This effect is useful in the design of transformers and eddy current probes⑧.

1.5.2 Mechanical Properties of Materials

The mechanical properties of a material are those ones that involve a reaction to an applied load. The mechanical properties of metals determine the range of usefulness of a material and establish the service life that can be expected. Mechanical properties are also used to help classify and identify material. The most common properties considered are strength, ductility, hardness, impact resistance and fracture toughness.

Most structural materials are anisotropic, which means that their material properties vary with orientation. The variation in properties can be due to directionality in the microstructure (texture) from forming or cold working operation, the controlled alignment of fiber reinforcement and a variety of other causes. Mechanical properties are generally specific to product form such as sheet, plate, extrusion, casting, forging, *etc*. Additionally, it is common to see mechanical property listed by the directional grain structure of the material. In products such as sheet and plate, the rolling direction is called the longitudinal direction, the width of the product is called the transverse direction, and the thickness is called the short transverse direction. The grain orientations in standard wrought forms of metallic products are shown the image.

The mechanical properties of a material are not constant and often change as a

Chapter 1 Introduction

function of temperature, rate of loading and other conditions. For example, temperatures below room temperature generally cause an increase in strength properties of metallic alloys; while ductility, fracture toughness and elongation usually decrease. Temperatures above room temperature usually cause a decrease in the strength properties of metallic alloys. Ductility may increase or decrease with increasing temperature depending on the same variables.

It should also be noted that there is often significant variability in the values obtained when measuring mechanical properties. Seemingly identical test specimen from the same lot of material will often produce considerable different results. Therefore, multiple tests are commonly conducted to determine mechanical properties and values reported can be an average value or calculated statistical minimum value. Also, a range of values is sometimes reported in order to show variability.

(1) Loading

The application of a force to an object is known as loading. Materials can be subjected to many different loading scenarios and a materials performance is dependant on the loading conditions. There are five fundamental loading conditions: tension, compression, bending, shear and torsion. Tension is the type of loading in which the two sections of material on either side of a plane tend to be pulled apart or elongated. Compression is the reverse of tensile loading and involves pressing the material together. Loading by bending involve applying a load in a manner that causes a material to curve and results in compressing the material on one side and stretching it on the other. Shear involves applying a load parallel to a plane which caused the material on one side of the plane to want to slide across the material on the other side of the plane. Torsion is the application of a force that causes twisting in a material.

If a material is subjected to a constant force, it is called static loading. If the loading of the material is not constant but instead fluctuates, it is called dynamic or cyclic loading. The way a material is loaded greatly affects its mechanical properties and largely determines how, or if, a component will fail; and whether it will show warning signs before failure actually occurs.

(2) Stress

The term stress(S) is used to express the loading in terms of force applied to a certain cross-sectional area of an object. From the perspective of loading, stress is the applied force or system of forces that tends to deform a body. From the perspective of what is happening within a material, stress is the internal distribution of forces within a body that balance and react to the loads applied to it. The stress distribution may or may not be uniform, depending on the nature of the loading condition. For example, a bar loaded in pure tension will essentially have a uniform tensile stress distribution. However, a bar loaded in bending will have a stress distribution that changes with distance perpendicular to the normal axis.

(3) Strain

Strain is the response of a system to an applied stress. When a material is loaded with a force, it produces a stress, which then causes a material to deform. Engineering strain is defined as the amount of deformation in the direction of the applied force divided by the initial length of the material. This results in a unitless number, although it is often left in the unsimplified form, such as inches per inch or meters per meter. For example, the strain in a bar that is being stretched in tension is the amount of elongation or change in length divided by its original length. As in the case of stress, the strain distribution may or may not be uniform in a complex structural element, depending on the nature of the loading condition.

If the stress is small, the material may only strain a small amount and the material will return to its original size after the stress is released. This is called elastic deformation, because of liking elastic, it returns to its unstressed state. Elastic deformation only occurs in a material when stresses are lower than a critical stress called the yield strength. If a material is loaded beyond it elastic limit, the material will remain in a deformed condition after the load is removed. This is called plastic deformation[2].

(4) Tensile properties

Tensile properties indicate how the material will react to forces being applied in tension. A tensile test is a fundamental mechanical test where a carefully prepared specimen is loaded in a very controlled manner while measuring the applied load and the elongation of the specimen over some distance. Tensile tests are used to determine the modulus of elasticity, elastic limit, elongation, proportional limit, reduction in area, tensile strength, yield point, yield strength and other tensile properties.

(5) Hardness

Hardness is the resistance of a material to localized deformation. The term can apply to deformation from indentation, scratching, cutting or bending. In metals, ceramics and most polymers, the deformation considered is plastic deformation of the surface. For elastomers and some polymers, hardness is defined at the resistance to elastic deformation of the surface. The lack of a fundamental definition indicates that hardness is not be a basic property of a material, but rather a composite one with contributions from the yield strength, work hardening, true tensile strength, modulus and others factors. Hardness measurements are widely used for the quality control of materials because they are quick and considered to be nondestructive tests when the marks or indentations produced by the test are in low stress areas.

(6) Toughness

The ability of a metal to deform plastically and to absorb energy in the process before fracture is termed toughness. The emphasis of this definition should be placed on the ability to absorb energy before fracture. Recall that ductility is a measure of how much something deforms plastically before fracture, but just because a material is ductile does not make it tough. The key to toughness is a good combination of strength and ductility. A material with high strength and high ductility will have more toughness than a material with low strength and high ductility. Therefore, one way to measure toughness is by calculating the area under the stress strain curve from a tensile test. This value is simply called "material toughness" and it has units of energy per volume. Material toughness equates to a slow absorption of energy by the material.

1.6 Materials Science

The concept of materials science arose from the necessity of acquiring knowledge of the fundamental laws, which determine properties. Materials science seeks to establish the relations existing between composition, atomic or molecular organization, the microstructure and the macroscopic properties of materials. Materials engineering concerned with manufacturing, transformation and shaping processes, complements materials science.

A fundamental knowledge of materials was not required when man contented himself with some clay, some wood and some wool to satisfy most of his necessities. The empirical approach and experience accumulated by the metalworkers and ceramists over thousands of years no longer suffice to satisfy contemporary needs and to meet the complex requirements of modern technology. A unified quantitative and fundamental approach to a description of the behavior of the engineering materials has become absolutely essential.

Materials science has a general character and a multi-disciplinary approach requiring the knowledge of chemists and physicists for basic sciences and those of the engineer (chemical, mechanical, electrical and civil) for applications and manufacturing. Materials science emerges as a coherent whole, coupled with materials engineering which has the objective the producing materials of well defined properties. Materials science treats the whole of materials (metals, ceramic and polymers) ina unified way with the same theoretical background and using the same experimental tools. As schematized in Figure 1.7, there are four main aspects materials science and technology: synthesis, manufacturing and processing, composition and structure, properties and performances. The behavior in manufacture and in use coupled with economic factors characterizes the performance of a material. Closely linked are four aspects of materials science. The materials is elaborated during synthesis (polymer) or manufacturing (metals, alloys, ceramics, *etc.*). Processing concerns the shaping of a material and the preparation of a finished object according to its behavior. For example, the production of a car body involves successively rolling of the sheet steel from a bar of steel, the stamping of the sheet steel to form the body and a series of finishing operation (painting, *etc.*).

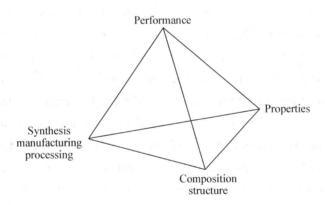

Figure 1.7 **The four basic aspects of materials science and technology**

To obtain optimal properties, it is essential to master the structure and composition of the material and consequently to have access to a series of sophisticated analysis techniques.

It is the numerous contributions of materials science and technology, which has completely remodeled the world, which supports us by freeing man of a huge number of constraints, linked to our environment. Our way of life has been radically transformed within a few decades largely due to the contributions of materials science and engineering which lead to the creation of the tools of the modern life: automobiles, aircraft, bridges, cable cars, computers, telecommunications equipment, satellites, *etc.*

1.7 Modern Materials' Needs

In spite of the tremendous progress that has been made in the discipline of materials science and engineering within the past few years, there stillremain technological challenges, including the development of even more sophisticated and specialized materials, as well as consideration of the environmental impact of materials production. Some comment is appropriate relative to these issues so as to round out this perspective.

Nuclear energy holds some promise, but the solutions to the many problems that remain will necessarily involve materials, from fuels to containment structures to facilities for the disposal of radioactive waste.

Significant quantities of energy are involved in transportation. Reducing the weight of transportation vehicles (automobiles, aircraft, trains, *etc.*), as well as increasing engine operating temperatures, will enhance fuel efficiency. New high-strength and low-density structural materials remain to be developed, as well as materials that have higher-temperature capabilities, for use in engine components.

Furthermore, there is a recognized need to find new, economical sources of energy, and to use the present more efficiently. Materials will undoubtedly play a significant role in these developments. For example, the direct conversion of solar into electrical energy has been demonstrated. Solar cells employ some rather complex and expensive materials. To ensure a viable technology, materials that are highly efficient in this conversion process yet less costly must be developed.

Furthermore, environmental quality depends on our ability to control air and water pollution. Pollution control techniques employ various materials. In addition, materials processing and refinement methods need to be improved so that they produce less environmental degradation, that is, less pollution and less despoilage of the landscape from the mining of raw materials. Also, in some materials manufacturing processes, toxic substances are produced, and the ecological impact of their disposal must be considered.

Many materials that we use are derived from resources that are nonrenewable, that is, not capable of being regenerated. These include polymers, for which the prime raw material is oil, and some metals. These nonrenewable resources are gradually becoming depleted, which necessitates: (1) the discovery ofadditional reserves, (2) the development of new materials having comparable properties with less adverse environmental impact, and/or (3) increased recycling efforts and the development of new recycling technologies. As a consequence of the economics of not only production but also environmental impact and ecological factors, it is becoming increasingly important to consider the "cradle-to-grave" life cycle of materials relative to the overall manufacturing process.

◆ Notes

① deep-seated 深层的；根深蒂固的
② state-of-the-art 最先进的；已经发展的；达到最高水准的
③ noise-canceling 噪声消除
④ top-down 自上而下的，由大到小的
⑤ bottom-up 自下向上的；从细节到总体的
⑥ close-packed 拥挤不堪的；装得满满的
⑦ Miller indices 密勒指数
⑧ eddy current probes 涡流探针

◆ Vocabulary

actuator ['æktjʊeɪtə] n. 操作者，开动者；驱使者，激励者
aerodynamic [ˌɛərəʊdaɪ'næmɪk] adj. 空气动力学的，[航]航空动力学的
alcohol ['ælkəhɒl] n. 酒精，乙醇
amorphous [ə'mɔːfəs] adj. 无定形的；无组织的；[物]非晶形的
anisotropy [ˌænaɪ'sɒtrəpɪ] n. [物]各向异性
antinoise [ˌæntɪ'nɔɪz] adj. 抗噪音的
benzene ['benziːn] n. [有化]苯
biological [ˌbaɪə'lɒdʒɪkəl] adj. 生物的；生物学的
biomaterial [ˌbaɪəmə'tɪrɪəl] adj. [材]生物材料
camcorders ['kæmkɔːdə] n. 摄录像机；便携式摄像机
carbide ['kɑːbaɪd] n. [无化]碳化物
cement [sɪ'ment] n. 水泥；接合剂
ceramics [sə'ræmɪks] n. 制陶术，制陶业（ceramic 的复数）
circuitry ['sɜːkɪtrɪ] n. 电路；电路系统；电路学

combustible [kəm'bʌstəbl] adj. 易燃的；燃烧性的
condense [kən'dens] vt. 使浓缩；使压缩 vi. 浓缩；凝结
corrosion [kə'rəʊʒən] n. 腐蚀；腐蚀产生的物质；衰败
crystalline ['krɪstəlaɪn] adj. 晶质的，结晶的；由结晶体组成的；结晶状的
crystallographic [ˌkrɪstələʊ'græfɪk] adj. 结晶的
degradation [ˌdegrə'deɪʃən] n. 退化；降格，降级；堕落
diamagnetic [ˌdaɪəmæg'netɪk] n. 反磁性体 adj. 反磁性的
ductility [dʌk'tɪlətɪ] n. 延性，延展性，韧性
dynamic [daɪ'næmɪk] n. 动态；动力 adj. 动力的；动力学的
elastic [ɪ'læstɪk] n. 松紧带；橡皮圈 adj. 有弹性的；灵活的；易伸缩的
elasticity [elæ'stɪsɪtɪ] n. 弹性，[物理学]弹性，弹力
elastomer [ɪ'læstəmə] n. [力]弹性体，

Chapter 1 Introduction

[高分子]高弹体
elongation [ˌiːlɒŋˈgeɪʃən] n. 伸长;伸长率;延伸率;延长
empirical [emˈpɪrɪkəl] adj. 经验主义的,完全根据经验的;实证的
ferromagnetic [ˌferəʊmægˈnetɪk] adj. 铁磁的,铁磁性,磁性铁的
fiberglass [ˈfaɪbəglɑːs] n. 玻璃纤维;玻璃丝
flammability [ˌflæməˈbɪlətɪ] n. 可燃性,易燃性
flexible [ˈfleksɪbl] adj. 灵活的;柔韧的;易弯曲的
formula [ˈfɔːmjʊlə] n. 公式,配方,数学公式
geometry [dʒɪˈamɪtrɪ] n. 几何,几何学,几何体
hardness [ˈhɑːdnəs] n. [物]硬度;坚硬
helicopter [ˈhelɪkaptə] n. 直升机,直升飞机,直升机
hexagonal [hekˈsægənəl] adj. 六边的,六角形的
hydroelectric [ˌhaɪdrəʊɪˈlektrɪk] adj. 水力发电的;水电治疗的
hydrogen [ˈhaɪdrədʒən] n. [化学]氢
implanted [ɪmˈplɑːntɪd] adj. 植入的
Inhomogeneity [ɪnˌhɒməʊdʒeˈniːətɪ] n. 不均一;不同类;不同质
inorganic [ɪnɔːˈgænɪk] adj. [无化]无机的;无生物的
insulative [ˈɪnsjʊleɪtɪv] adj. 绝缘的;隔音的
insulator [ˈɪnsjʊleɪtə] n. [物]绝缘体;从事绝缘工作的工人
intermediate [ˌɪntəˈmiːdɪət] n. [化学]中间物;媒介 adj. 中间的,中级的 vi. 起媒介作用

interstice [ɪnˈtɜːstɪs] n. [建]裂缝;空隙
interstitial [ˌɪntəˈstɪʃəl] n. 填隙原子;节间 adj. 间质的;空隙的;填隙的
longitudinal [ˌlɒndʒɪˈtjuːdɪ(ə)l] adj. 长度的,纵向的;经线的
lustrous [ˈlʌstrəs] adj. 有光泽的;光辉的
magnetorheological
magnetostrictive [mægˌniːtəʊˈstrɪktɪv] adj. [物]磁致伸缩的
mercury [ˈmɜːkjəri] n. [化]汞,水银
microvoids [ˌmaɪkrəʊˈvɒɪd] n. 微孔;微空洞
modulus [ˈmɒdjʊləs] n. 系数,模数,模量
molecular [məˈlekjʊlə] adj. [化学]分子的;由分子组成的
molecules [ˈmɒlɪkjuːls] n. [化学]分子,微粒;[化学]摩尔(molecule 的复数)
monocrystal [ˈmɒnəˌkrɪstəl] n. 单晶体 adj. 单晶的
multidisciplinary [ˌmʌltɪdɪsəˈplɪnəri] adj. 有关各种学问的
multiple [ˈmʌltɪpl] n. 倍数;[电]并联 adj. 多重的;多样的;许多的
nitride [ˈnaɪtraɪd] n. [无化]氮化物,氮化物,氮化
nondestructive [ˌnɒndɪˈstrʌktɪv] adj. 无损的;非破坏性的
nonmetallic [ˌnɒnmɪˈtælɪk] n. 非金属物质 adj. 非金属的
nonrenewable [ˌnɒnrɪˈnjuːəbl] adj. 不可再生的;不可更新的
octahedral [ˌɒktəˈhedrəl] adj. 八面体的;有八面的
organic [ɔːˈgænɪk] adj. [有化]有机的;组织的;器官的;根本的

organisms ['ɔrgən,ɪzəm] n. [生物]生物体(organism 的复数);[生物]有机体
parallelepiped [,pærəlelə'paɪped] n. 平行六面体
paramagnetic [,pærəmæg'netɪk] n. 顺磁物质;顺磁体 adj. 顺磁性的;常磁性的
permeability [pɜːmɪə'bɪlɪtɪ] n. 透磁率,导磁系数
perpendicular [,pɜːp(ə)n'dɪkjʊlə] n. 垂线;垂直的位置 adj. 垂直的,正交的;直立的
piezoelectric [piːˌeɪzəʊɪ'lektrɪk] adj. [电]压电的
plasma ['plæzmə] n. [等离子]等离子体
pliable ['plaɪəbl] adj. 柔韧的;柔软的;易曲折的
plurality [plʊə'rælɪtɪ] n. 多数;复数
polycrystal [,pɒlɪ'krɪstl] n. [晶体]多晶体
polymer ['pɒlɪmə] n. [高分子]聚合物
polymeric [,pɒlɪ'merɪk] adj. 聚合的;聚合体的
pottery ['pɒtərɪ] n. 陶器;陶器厂;陶器制造术
predetermined [,priːdɪ'tɜːmɪnd] adj. 预定的,预先决定,预先约定的
proportionality [prəʊˌpɔːʃə'nælɪtɪ] n. 相称;均衡;比例性
radioactive [,reɪdɪəʊ'æktɪv] adj. [核]放射性的;有辐射的
remodeled [riː'mɒdld] adj. 改造的;改制的
resistivity [,rɪzɪ'stɪvɪtɪ] n. [电]电阻率;电阻系数
semiconductor [,semɪkən'dʌktə] n. [电子][物]半导体
shear [ʃɪə] n. [力]切变;修剪;大剪刀 vi. 剪;剪切;修剪 vt. 剪;修剪
solidify [sə'lɪdɪfaɪ] vt. 凝固 vi. 凝固
sophisticated [sə'fɪstɪkeɪtɪd] adj. 复杂的;精致的;久经世故的;富有经验的
spatial ['speɪʃəl] adj. 空间的;存在于空间的;受空间条件限制的
specimen ['spesɪmɪn] n. 样品,样本;标本
statistic [stə'tɪstɪk] n. 统计数值 adj. 统计的,统计学的
sublime [sə'blaɪm] vt. 使……纯化;使……升华
submicroscopic [,sʌbmaɪkrə'skɒpɪk] adj. 亚微观的
symbolize ['sɪmbəlaɪz] vt. 象征;用符号表现 vi. 采用象征;使用符号;作为……的象征
symmetry ['sɪmɪtrɪ] n. 对称(性);整齐,匀称
tetrahedral [,tetrə'hiːdrəl] adj. 四面体的;有四面的
tissue ['tɪʃuː] n. 组织
torsion ['tɔːʃən] n. 扭转,扭曲;转矩,[力]扭力
toxic ['tɒksɪk] adj. 有毒的;中毒的
transition [træn'zɪʃən] n. 过渡;转变;[分子生物]转换;变调
transparent [træn'spærənt] adj. 透明的;显然的;坦率的;易懂的
transverse [trænz'vɜːs] n. 横断面;贯轴;横肌 adj. 横向的;横断的;贯轴的
turbidity [tɜː'bɪdətɪ] n. [分化]浊度;浑浊;混浊度
twist [twɪst] n. 扭曲,拧;扭伤 vt. 捻;拧;扭伤 vi. 扭动;弯曲
vaporize ['veɪpəraɪz] vt. 使……蒸发 vi. 蒸发

Chapter 1　Introduction

Exercises

1. Translate the following Chinese phrases into English

(1) 材料科学　　　　(2) 固体材料　　　　(3) 智能材料

(4) 碳纳米管　　　　(5) 相对密度　　　　(6) 宏(或微)观结构

(7) 玻璃纤维　　　　(8) 陶瓷材料　　　　(9) 物理(或化学)反应

(10) 有色金属　　　(11) 高科技　　　　　(12) 形状记忆合金

(13) 准各向同性　　(14) 比重　　　　　　(15) 室温

(16) 截面的(断面的)

2. Translate the following English phrases into Chinese

(1) mechanical property　　　　　　(2) advanced materials

(3) high-performance material(s)　　(4) nanoengineering material(s)

(5) phase transition temperature　　(6) yield strength

(7) crystal lattice　　　　　　　　　(8) crystalline (or amorphous) state

(9) body-centred cubic lattice　　　 (10) magnetic permeability

(11) elastic deformation　　　　　　(12) fresh water

3. Translate the following Chinese sentences into English

(1) 在常温常压下,成千上万种固体物质可分为金属、陶瓷、半导体、复合材料和高分子材料。

(2) 直到最近,科学家才终于了解到材料的结构要素与其特性之间的关系。

(3) 任何物质,无论是固体、液体还是气体,都是由原子组成的。

(4) 大部分高分子材料具有差的导电性。

(5) 纳米材料是指三维空间中至少有一维尺寸处于纳米量级(1～100 nm)的材料。

4. Translate the following English sentences into Chinese

(1) Composite materials are structured materials composed of two or more macroscopic phases.

(2) It is well known that one of the subatomic particles of an atom is the electron.

(3) The mechanical properties of a material are not constant and often change as a function of temperature, rate of loading and other conditions.

(4) It should be noted that the dividing line among the various types of plastics is not based on material but rather on their properties and application.

(5) For example, magnesium with an atomic number of 12, has two electrons in the inner shell, eight in the second sell and two in the outer shell.

5. Translate the following Chinese essay into English

当材料结构中的空间维度减小,或者纳米粒子在一个特定的晶体方向上受到限制时,通常会导致材料在该晶向上的物理性能发生变化。因而,可依据材料处于纳米范围内的空间维度的数量,对纳米材料和纳米体系进行分类。

(1) 零维(0D)材料

零维材料是指材料的空间三维结构尺度均处于纳米量级,也就是说,材料的三维尺寸均被限制在纳米尺度内。该系统的材料包括纳米粒子和纳米晶等。

(2) 一维(1D)材料

一维材料是指材料的空间结构尺度中,有二维处于纳米量级,也就是说,其三维中的二维被限制在纳米量级内。该体系包括纳米线、纳米棒、纳米丝和纳米管等。

(3) 二维(2D)材料

二维材料是指在材料的空间结构尺度中,有一维处于纳米量级,也就是说,其三维中的一维被限制在纳米量级内。该体系包括超薄膜、多层膜、薄膜、表面涂层和超晶格等。

6. Translate the following English essay into Chinese

Materials are the matter of the universe. These substances have properties that make useful in structures, machines, devices, products and systems. The term properties describe behavior of materials when subjected to some external force or condition. For example, the tensile strength of a metal is a measure of the material's resistance to a pulling force. The family of materials consists of four main groups of materials: metals (e.g., steel), polymers (e.g., plastics), ceramics (e.g., porcelain) and composites (e.g., glass-reinforced plastics). The materials in each group have similar properties and/or structures.

扫一扫,查看更多资料

Part I
Foundation of Phase Change of Materials

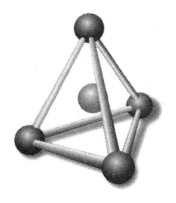

Chapter 2 Phase Diagrams

2.1 Introduction

The study of phase relationships plays an important and vital role in the better understanding of properties of materials. Much of the information about the control of microstructure or phase structure of a particular alloy system is properly displayed in what is called a phase diagram, also called as an equilibrium or constitutional diagram. And also, the understanding of phase diagrams for alloy systems is extremely important because there is a strong correlation between microstructure and mechanical properties, and the development of microstructure of an alloy is related to the characteristics of its phase diagram. Especially the equilibrium phase diagrams are a convenient and concise way of representing the most stable relationships between phases in alloy systems. For an alloy of specified composition and at a known temperature, the phases present, their compositions, and relative amounts under equilibrium conditions may be determined by phase diagrams. In addition, phase diagrams provide valuable information about melting, casting, crystallization, and other phenomena.

This chapter presents and discusses the following topics:

(1) Terminology associated with phase diagrams and phase transformations.

(2) The interpretation of phase diagrams.

(3) Some of the common and relatively simple binary phase diagrams, including that for the iron-carbon system[①].

2.2 Definitions and Basic Concepts

It is necessary to establish a foundation of definitions and basic concepts relating to alloys, phases, and equilibrium before delving into the interpretation and utilization of phase diagrams. The term component is frequently used in this discussion; components are pure metals and/or compounds of which an alloy is composed. For example, in a copper-zinc brass[②], the components are Cu and Zn. Solute and solvent are also common terms. Another term used in this context is system, which has two meanings. First, "system" may refer to a specific body of material under consideration (e.g., a ladle of molten steel). Or, it may relate to the series of possible alloys consisting of the same components, but without regard to alloy composition (e.g., the iron-carbon system).

A solid solution forms when, as the solute atoms are added to the host material, the crystal structure is maintained, and no new structures are formed. Perhaps it is useful to draw an analogy with a liquid solution. If two liquids, soluble in each other (such as water and alcohol) are combined, a liquid solution is produced as the molecules intermix, and its composition is homogeneous throughout. A solid solution is also compositionally homogeneous; the impurity atoms are randomly and uniformly dispersed within the solid. By way of review, a solid solution consists of atoms of at least two different types; the solute atoms occupy either substitutional or interstitial positions in the solvent lattice, and the crystal structure of the solvent is maintained.

2.3 Solubility Limit

For many alloy systems and at some specific temperature, there is a maximum concentration of solute atoms that may dissolve in the solvent to form a solid solution; this is called a solubility limit. The addition of solute in excess of this solubility limit results in the formation of another solid solution or compound that has a distinctly different composition. To illustrate this concept, consider the sugar-water[③] ($C_{12}H_{22}O_{11}$—H_2O) system. Initially, as sugar is added to water, a sugar-water solution or syrup forms. As more sugar is introduced, the solution becomes more concentrated, until the solubility limit is reached, or the solution becomes saturated with sugar. At this time the solution is not capable of dissolving any more sugar, and further additions simply settle to the bottom of the container. Thus, the system now consists of two separate substances: a sugar-water syrup liquid solution and solid crystals of un-dissolved sugar.

This solubility limit of sugar m water depends on the temperature of the water and may be represented in graphical form on a plot of temperature along the ordinate and composition (in weight percent sugar) along the abscissa, as shown in

Figure 2.1. Along the composition axis, increasing sugar concentration is from left to right, and percentage of water is read from right to left. Since only two components are involved (sugar and water), the sum of the concentrations at any composition will equal 100 *wt.* %. The solubility limit is represented as the nearly vertical line in the figure, for compositions and temperatures to the left of the solubility line, only the syrup liquid solution exists; to the right of the line, syrup and solid sugar coexist. The solubility limit at some temperature is the composition that corresponds to the intersection of the given temperature coordinate and the solubility limit line. For example, at 20 ℃ the maximum solubility of sugar in water is 65 *wt.* %. As indicated in Figure 2.1, the solubility limit increases slightly with rising temperature.

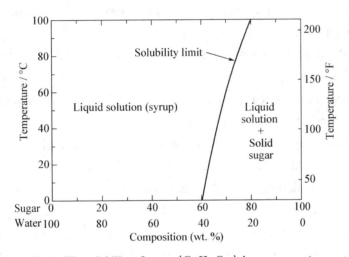

Figure 2.1 The solubility of sugar ($C_{12}H_{22}O_{11}$) in a sugar-water syrup

2.4 Phases

Also critical to the understanding of phase diagrams is the concept of a phase. A phase may be defined as a homogeneous portion of a system that has uniform physical and chemical characteristics. Every pure material is considered to be a phase; so also is every solid, liquid, and gaseous solution. For example, the sugar-water syrup solution just discussed is one phase, and solid sugar is the other. Each has different physical properties (one is a liquid, the other is a solid); furthermore each is different chemically (*i.e.*, has a different chemical composition); one is virtually pure sugar; the other is a solution of H_2O and $C_{12}H_{22}O_{11}$. If more than one phase is present in a given system, each will have its own distinct properties. And a boundary separating the phases will exist across which there will be a discontinuous and abrupt change in physical and/or chemical characteristics. When two phases are present in a system, it is not necessary that there be a difference in both physical and chemical properties; a disparity in one or the other set of properties is sufficient. When water and ice are present in a container, two separate phases exist; they are physically dissimilar (one is a solid, the other is a liquid) but identical in chemical makeup. Also, when a substance can exist in two or more polymorphic forms (e.g., having both FCC and BCC structures), each of these structures is a separate phase because their respective physical characteristics differ.

Sometimes, a single-phase system is termed "homogeneous". Systems composed of two or more phases are termed "mixtures" or "heterogeneous systems[④]". Most metallic alloys, and, for that matter, ceramic, polymeric, and composite systems are heterogeneous. Ordinarily, the phases interact in such a way that the property combination of the multiphase system is different from, and more attractive than, either of the individual phases.

Many times, the physical properties, in particular, the mechanical behavior of a material depend on the microstructure. Microstructure is subject to direct microscopic observation, using optical or electron microscopes. In metal alloys, microstructure is characterized by the number of phases present, their proportions, and the manner in which they are distributed or arranged. The microstructure of an alloy depends on such variables as the alloying elements present, their concentrations, and the heat treatment of the alloy (i.e., the temperature, the heating time at temperature, and the rate of cooling to room temperature).

As well-known, the procedure of specimen preparation for microscopic examination is very common. After appropriate polishing and etching, the different phases may be distinguished by their appearance. For example, for a two-phase alloy, one phase may appear light, and the other phase dark. When only a single phase or solid solution is present, the texture will be uniform, except for grain boundaries that may be revealed.

2.5 Phase Equilibrium

Equilibrium is another essential concept. It is best described in terms of a thermodynamic quantity called the free energy. In brief, free energy is a function of the internal energy of a system, and also the randomness or disorder of the atoms or molecules (or entropy). A system is at equilibrium if its free energy is at a minimum under some specified combination of temperature, pressure, and composition. In a macroscopic sense, this means that the characteristics of the system do not change with time but persist indefinitely: that is, the system is stable. A change in temperature, pressure, and/or composition for a system in equilibrium will result in an increase in the free energy and in a possible spontaneous change to another state whereby the free energy is lowered.

The term phase equilibrium, often used in the context of this discussion, refers to equilibrium as it applies to systems in which more than one phase may exist. Phase equilibrium is reflected by constancy with time in the phase characteristics of a system. Perhaps an example best illustrates this concept. Suppose that a sugar-water syrup is contained in a closed vessel and the solution is in contact with solid sugar at 20 ℃. If the system is at equilibrium, the composition of the syrup is 65 *wt*. % $C_{12}H_{22}O_{11}$ - 35 *wt*. % H_2O (Figure 2.1), and the amounts and compositions of the syrup and solid sugar will remain constant with time. If the temperature of the system is suddenly raised to 100 ℃, this equilibrium or balance is temporarily upset in that the solubility limit has been increased to 80 *wt*. % $C_{12}H_{22}O_{11}$ (Figure 2.1). Thus, some of the solid sugar will go into solution in the syrup. This will continue until the new equilibrium syrup concentration is established at the higher temperature.

This sugar-syrup example has illustrated the principle of phase equilibrium using a liquid-solid system. In many metallurgical and materials systems of interest, phase equilibrium involves just solid phases. In this regard the state of the system is reflected in the characteristics of the microstructure, which necessarily include not only the phases present and their compositions but, in addition, the relative phase amounts and their spatial arrangement or distribution.

Free energy considerations and diagrams similar to Figure 2.1 provide information about the equilibrium characteristics of a particular system, which is important; but they do not indicate the time period necessary for the attainment of a new equilibrium state. It is often the case, especially in solid systems, that a state of equilibrium is never completely achieved because the rate of approach to equilibrium is extremely slow; such a system is said to be in a non-equilibrium or metastable state. A metastable state or microstructure may persist indefinitely, experiencing only extremely slight and almost imperceptible changes as time progresses. Often, metastable structures are of more practical significance than equilibrium ones.

Thus not only is an understanding of equilibrium states and structures important, but the speed or rate at which they are established and, in addition, the factors that affect the rate must be considered.

2.6 Equilibrium Phase Diagrams

Much of the information about the control of microstructure or phase structure of a particular alloy system is conveniently and concisely displayed in what is called a phase diagram, also often termed an equilibrium or constitutional diagram. Many microstructures develop from phase transformations, the changes that occur between phases when the temperature is altered (ordinarily upon cooling). This may involve the transition from one phase to another or the appearance or disappearance of a phase. Phase diagrams are helpful in predicting phase transformations and the resulting microstructures, which may have equilibrium or non-equilibrium character.

Equilibrium phase diagrams represent the relationships between temperature and the compositions and the quantities of phases at equilibrium. There are several different varieties; but in the present discussion, temperature and composition are the variable parameters, for binary alloys. A binary alloy is one that contains two components. If more than two components are present, phase diagrams become extremely complicated and difficult to represent. The principles of microstructural control with the aid of phase diagrams can be illustrated with binary alloys even though, in reality, most alloys contain more than two components. External pressure is also a parameter that influences the phase structure. However, in practicality, pressure remains virtually constant in most applications; thus, the phase diagrams presented here are for a constant pressure of one atmosphere (1 atm).

Possibly the easiest type of binary phase diagram to understand and interpret is that which is characterized by the copper-nickel system (Figure 2.2a). Temperature is plotted along the ordinate, and the abscissa represents the composition of the alloy, in weight percent (bottom) and atom percent (top) of nickel.

The composition ranges from 0 $wt.$ % Ni (100 $wt.$ % Cu) on the left horizontal extremity to 100 $wt.$ % Ni (0 $wt.$ % Cu) on the right. Three different phase regions, or fields, appear on the diagram, an alpha (α) field, a liquid (L) field, and a two-phase $\alpha + L$ field. Each region is defined by the phase or phases that exist over the range of temperatures and compositions delimited by the phase boundary lines.

The liquid L is a homogeneous liquid solution composed of both copper and nickel. The α phase is a substitutional solid solution consisting of both Cu and Ni atoms, and having an FCC crystal structure. At temperatures below about 1 080 ℃, copper and nickel are mutually soluble in each other in the solid state for all compositions. This complete solubility is explained by the fact that both Cu and Ni have the same crystal structure (FCC), nearly identical atomic radii and electronegativities. The copper-nickel system is termed isomorphous because of this complete liquid and solid solubility of the two components.

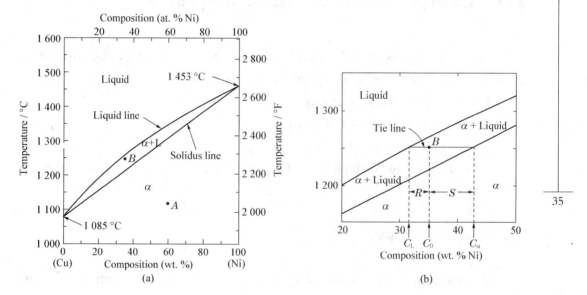

Figure 2.2 The copper-nickel local phase diagram (a) and the copper-nickel local phase diagram (b)

A couple of comments are in order regarding nomenclature. First, for metallic alloys, solid solutions are commonly designated by lowercase Greek letters (α, β, γ, etc.). Furthermore, with regard to phase boundaries, the line separating the L and $\alpha + L$ phase fields is termed the liquidus line, as indicated in Figure 2.2a. The liquid phase is present at all temperatures and compositions above this line. The solidus line is located between the α and $\alpha + L$ regions, below which only the solid α phase exists. For Figure 2.2a, the solidus and liquidus lines intersect at the two composition extremities; these correspond to the melting temperatures of the pure components. For example, the melting temperatures of pure copper and nickel are 1 085 and 1 455 ℃, respectively. Heating pure copper corresponds to moving vertically up the left-hand temperature axis. Copper remains solid until its melting temperature is reached. The solid-to-liquid transformation takes place at the melting temperature, and no further heating is possible until this transformation has been completed.

For any composition other than pure components, this melting phenomenon will occur over the range of temperatures between the solidus and liquidus lines, both solid and liquid phases will be in equilibrium within this temperature range. For example, upon heating an alloy of composition 50 *wt*. % Ni – 50 *wt*. % Cu (Figure 2.2a), melting begins at approximately 1 280 ℃ (2 340 ℉); the amount of liquid phase continuously increases with temperature until about 1 320 ℃ (2 410 ℉), at which the alloy is completely liquid.

2.7　Interpretation of Phase Diagrams

For a binary system of known composition and temperature that is at equilibrium, at least three kinds of information are available:
(1) The phases that are present.
(2) The compositions of these phases.
(3) The percentages or fractions of the phases.

The procedures for making these determinations will be demonstrated using the copper-nickel system.

2.7.1　Phases Present

The establishment of what phases are present is relatively simple. One just locates the temperature-composition point on the diagram and notes the phase(s) with which the corresponding phase field is labeled. For example, an alloy of composition 60 *wt*. % Ni – 40 *wt*. Cu at 1 100 ℃ would be located at point A in Figure 2.2a; since this is within the α region, only the single α phase will be present. On the other hand, a 35 *wt*. % Ni – 65 *wt*. % Cu alloy at 1 250 ℃ (point B) will consist of both α and liquid phases at equilibrium.

2.7.2　Determination of Phase Compositions

The first step in the determination of phase compositions (in terms of the concentrations of the components) is to locate the temperature-composition point on the phase diagram. Different methods are used for single-and two-phase regions. If only one phase is present, the procedure is trivial: the composition of this phase is simply the same as the overall composition of the alloy. For example, consider the 60 *wt*. % Ni – 40 *wt*. % Cu alloy at 1 100 ℃ (point A, Figure 2.2a). At this composition and temperature, only the α phase is present, having a composition of 60 *wt*. % Ni – 40 *wt*. % Cu.

For an alloy having composition and temperature located in a two-phase region, the situation is more complicated. In all two-phase regions (and in two-phase regions only), one may imagine a series of horizontal lines, one at every temperature; each of these is known as a tie line, or sometimes as an isotherm. These tie lines extend across the two-phase region and terminate at the phase boundary lines on either side. To compute the equilibrium concentrations of the two phases, the following procedure is used:

(1) A tie line is constructed across the two-phase region at the temperature of the alloy.

(2) The intersections of the tie line and the phase boundaries on either side are noted.

(3) Perpendiculars are dropped from these intersections to the horizontal composition axis from which the composition of each of the respective phases is read.

For example, consider again the 35 *wt*. % Ni − 65 *wt*. % Cu alloy at 1 250 °C, located at point *B* in Figure 2.2b and lying within the α + *L* region. Thus, the problem is to determine the composition (in *wt*. % Ni and Cu) for both the α and liquid phases. The tie line has been constructed across the α + *L* phase region, as shown in Figure 2.2b. The perpendicular from the intersection of the tie line with the liquidus boundary meets the composition axis at 32 *wt*. % Ni − 68 *wt*. % Cu, which is the composition of the liquid phase, C_L. Likewise, for the solidus-tie line intersection, we find a composition for the a solid-solution phase, C_α, of 43 *wt*. % Ni − 57 *wt*. % Cu.

2.7.3 Determination of Phase Amounts

The relative amounts (as fraction or as percentage) of the phases present at equilibrium may also be computed with the aid of phase diagrams. Again, the single-and two-phase situations must be treated separately. The solution is obvious in the single-phase region: Since only one phase is present, the alloy is composed entirely of that phase; that is, the phase fraction is 1.0 or, alternatively, the percentage is 100%. From the previous example for the 60 *wt*. % Ni − 40 *wt*. % Cu alloy at 1 100 °C (point *A* in Figure 2.2a), only the α phase is present; hence, the alloy is completely or 100% α.

If the composition and temperature position is located within a two-phase region, things are more complex. The tie line must be utilized in conjunction with a procedure that is often called the lever rule, which is applied as follows:

(1) The tie line is constructed across the two-phase region at the temperature of

the alloy.

(2) The overall alloy composition is located on the tie line.

(3) The fraction of one phase is computed by taking the length of tie line from the overall alloy composition to the phase boundary for the other phase, and dividing by the total tie line length.

(4) The fraction of the other phase is determined in the same manner.

(5) If phase percentages are desired, each phase fraction is multiplied by 100. When the composition axis is scaled in weight percent, the phase fractions computed using the lever rule are mass fraction the mass (or weight) of a specific phase divided by the total alloy mass (or weight). The mass of each phase is computed from the product of each phase fraction and the total alloy mass.

In the employment of the lever rule, tie line segment lengths may be determined either by direct measurement from the phase diagram using a linear scale, preferably graduated in millimeters, or by subtracting compositions as taken from the composition axis.

Consider again the example shown in Figure 2.2b, in which at 1 250 °C both α and liquid phases are present for a 35 $wt.$ % Ni - 65 $wt.$ % Cu alloy. The problem is to compute the fraction of each of the α and liquid phases. The tie line has been constructed that was used for the determination of α and L phase compositions. Let the overall alloy composition be located along the tie line and denoted as C_0, and mass fractions be represented by W_L and W_α for the respective phases. From the lever rule, W_L may be computed according to

$$W_L = \frac{S}{R+S} \tag{2-1}$$

or, by subtracting compositions,

$$W_L = \frac{C_\alpha - C_0}{C_\alpha - C_L} \tag{2-2}$$

Composition need be specified in terms of only one of the constituents for a binary alloy; for the computation above, weight percent nickel will be used (i.e., $C_0 = 35$ $wt.$ % Ni, $C_\alpha = 43$ $wt.$ % Ni, and $C_L = 32$ $wt.$ % Ni), and

$$W_L = \frac{43 - 35}{43 - 31.5} = 0.73 \tag{2-3}$$

Similarly, for the α phase,

$$W_\alpha = \frac{S}{R+S} = \frac{C_0 - C_L}{C_\alpha - C_L} = \frac{35 - 32}{43 - 32} = 0.27 \tag{2-4}$$

Of course, identical answers are obtained if compositions are expressed in weight percent copper instead of nickel.

Thus, the lever rule may be employed to determine the relative amounts or fractions of phases in any two-phase region for a binary alloy if the temperature and composition are known and if equilibrium has been established. Its derivation is presented as an example problem.

It is easy to confuse the foregoing procedures for the determination of phase compositions and fractional phase amounts; thus, a brief summary is warranted. Compositions of phases are expressed in terms of weight percents of the components (e.g., wt. % Cu, wt. % Ni). For any alloy consisting of a single phase, the composition of that phase is the same as the total alloy composition. If two phases are present, the tie line must be employed, the extremities of which determine the compositions of the respective phases. With regard to fractional phase amounts (e.g., mass fraction of α or liquid phase), when a single phase exists, the alloy is completely that phase. For a two-phase alloy, on the other hand, the lever rule is utilized, in which a ratio of tie line segment lengths is taken.

2.7.4 Binary Eutectic Systems

Another type of common and relatively simple phase diagram found for binary alloys is shown in Figure 2.3 for the copper-silver system[5]; this is known as a binary eutectic phase diagram. A number of features of this phase diagram are important and worth noting. First of all, three single-phase regions are found on the diagram: α, β, and liquid. The α phase is a solid solution rich in copper and it has silver as the solute component and an FCC crystal structure. The β phase solid solution also has an FCC structure, but copper is the solute. Technically, pure copper and pure silver are considered to be α and β phases, respectively.

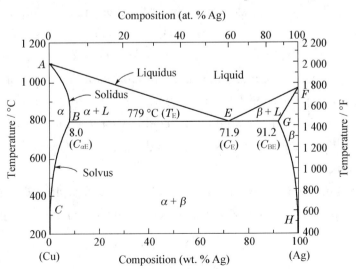

Figure 2.3 The copper-silver phase diagram

Thus, the solubility in each of these solid phases is limited, in that at any temperature below line BEG only a limited concentration of silver will dissolve in copper (for the α phase), and similarly for copper in silver (for the β phase). The solubility limit for the a phase corresponds to the boundary line, labeled CBA, between the α/(α + β) and α/(α + L) phase regions; it increases with temperature to a maximum [7.9 *wt*. % Ag at 780 ℃ (1 436 ℉)] *at point B, and decreases back to zero at the melting temperature of pure copper, point A* [1 085 ℃ (1 985 ℉)]. At temperatures below 780 ℃ (1 436 ℉), the solid solubility limit line separating the α and α + β phase regions is termed a solvus line; the boundary AB between the α and α + L fields is the solidus line, as indicated in Figure 2.3. For the β phase, both solvus and solidus lines also exist, HG and GF, respectively, as shown. The maximum solubility of copper in the β phase, point G (8.8 *wt*. % Cu), also occurs at 780 ℃ (1 436 ℉). This horizontal line BEG, which is parallel to the composition axis and extends between these maximum solubility positions, may also be considered to be a solidus line; it represents the lowest temperature at which a liquid phase may exist for any copper-silver alloy that is at equilibrium.

There are also three two-phase regions found for the copper-silver system (Figure 2.3): α + L, β + L, and α + β. The α and β phase solid solutions coexist for all compositions and temperatures within the α + β phase field; the α + liquid and β + liquid phases also coexist in their respective phase regions. Furthermore, compositions and relative amounts for the phases may be determined using tie lines and the lever rule as outlined in the preceding section.

As silver is added to copper, the temperature at which the alloys become totally liquid decreases along the liquidus line, line AE; thus, the melting temperature of copper is lowered by silver additions. The same may be said for silver, the introduction of copper reduces the temperature of complete melting along the other liquidus line, FE. These liquidus lines meet at the point E on the phase diagram, through which also passes the horizontal isotherm line BEG. Point E is called an invariant point, which is designated by the composition C_E and temperature T_E; for the copper-silver system, the values of C_E and T_E are 71.9 *wt*. % Ag and 780 ℃ (1 436 ℉), respectively.

An important reaction occurs for an alloy of composition C_E as it changes temperature in passing through T_E; this reaction may be written as follows:

$$L(C_E) \underset{\text{heating}}{\overset{\text{cooling}}{\rightleftharpoons}} \alpha(C_{\alpha E}) + \beta(C_{\beta E}) \qquad (2-5)$$

Or, upon cooling, a liquid phase is transformed into the two solid α and β phases at the temperature T_E; the opposite reaction occurs upon heating. This is called a eutectic reaction (eutectic means easily melted), and C_E and T_E represent the eutectic composition and temperature, respectively; and are the respective compositions of the α and β phases at T_E. Thus, for the copper-silver system, the eutectic reaction may be written as follows:

$$L(71.9 \text{ wt. \% Ag}) \underset{\text{heating}}{\overset{\text{cooling}}{\rightleftharpoons}} \alpha(8.0 \text{ wt. \% Ag}) + \beta(91.2 \text{ wt. \% Ag}) \qquad (2-6)$$

Often, the horizontal solidus line at T_E is called the eutectic isotherm.

The eutectic reaction, upon cooling, is similar to solidification for pure components m that the reaction proceeds to completion at a constant temperature, or isothermally, at T_E. However, the solid product of eutectic solidification is always two solid phases, whereas for a pure component only a single phase forms. Because of this eutectic reaction, phase diagrams similar to that in Figure 2.3 are termed eutectic phase diagrams; components exhibiting this behavior comprise a eutectic system.

In the construction of binary phase diagrams, it is important to understand that one or at most two phases may be in equilibrium within a phase field. This holds true for the phase diagrams in Figures 2.2a and 2.3. For a eutectic system, three phases (α, β, and L) may be in equilibrium, but only at points along the eutectic isotherm. Another general rule is that single-phase regions are always separated from each other by a two-phase region that consists of the two single phases that it separates. For example, the $\alpha + \beta$ field is situated between the α and β single-phase regions in Figure 2.3.

2.7.5 Gibbs Phase Rule

The construction of phase diagrams as well as some of the principles governing the conditions for phase equilibrium is dictated by laws of thermodynamics. One of these is the Gibbs phase rule, proposed by the nineteenth-century physicist J. Willard Gibbs. This rule represents a criterion for the number of phases that will coexist within a system at equilibrium, and is expressed by the simple equation

$$P + F = C + N \qquad (2-7)$$

where P is the number of phases present (the phase concept is discussed in Section 2.4). The parameter F is termed the number of degrees of freedom or the number of externally controlled variables (e.g., temperature, pressure, composition) which must be specified to completely define the state of the system. Or, expressed another way, F is the number of these variables that can be changed independently without altering the number of phases that coexist at equilibrium. The parameter C

in Equation (2-7) represents the number of components in the system. Components are normally elements or stable compounds and, in the case of phase diagrams, are the materials at the two extremities of the horizontal compositional axis (e.g., H_2O and $C_{12}H_{22}O_{11}$, and Cu and Ni for the phase diagrams shown in Figures 2.1 and 2.2a, respectively). Finally, N in Equation (2-7) is the number of non-compositional variables (e.g., temperature and pressure).

Let us demonstrate the phase rule by applying it to binary temperature-composition phase diagrams, specifically the copper-silver system, Figure 2.3. Since pressure is constant (1atm), the parameter N is 1, temperature is the only noncom-positional variable. Equation (2-7) now takes the form

$$P + F = C + 1 \qquad (2-8)$$

Furthermore, the number of components C is 2 (viz. Cu and Ag), and

$$P + F = 2 + 1 = 3 \qquad (2-9)$$

or

$$F = 3 - P \qquad (2-10)$$

Consider the case of single-phase fields on the phase diagram (e.g., α, β, and liquid regions).

Since only one phase is present, $P = 1$ and

$$F = 3 - P = 3 - 1 = 2 \qquad (2-11)$$

This means that to completely describe the characteristics of any alloy that exists within one of these phase fields, we must specify two parameters; these are composition and temperature, which locate, respectively, the horizontal and vertical positions of the alloy on the phase diagram.

For the situation wherein two phases coexist, for example, $\alpha + \beta$, $\beta + L$, and $\alpha + \beta$ phase regions, Figure 2.3, the phase rule stipulates that we have but one degree of freedom since

$$F = 3 - P = 3 - 2 = 1 \qquad (2-12)$$

Thus, it is necessary to specify either temperature or the composition of one of the phases to completely define the system. For example, suppose that we decide to specify temperature for the $\alpha + L$ phase region, say T_1 in Figure 2.4. The composition of α and liquid phases (C_α and C_L) are thus dictated by the extremities of the tie line constructed at T_1 across $\alpha + L$ field. It should be noted that only the nature of the phases is important in this treatment and not the relative phase amounts. This is to say that the overall alloy composition could lie anywhere along this tie line constructed at temperature T_1 and still give C_α and C_L compositions for the respective a and liquid phases.

The second alternative is to stipulate the composition of one of the phases for this two-phase situation, which thereby fixes completely the state of the system. For example, if we specified C_α as the composition of α phase that is in equilibrium with the liquid (Figure 2.4), then both the temperature of the alloy (T_1) and the composition of the liquid phase (C_L) are established, again by the tie line drawn across $\alpha + L$ phase field so as to give this C_α composition.

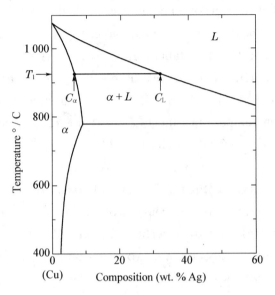

Figure 2.4 Copper-silver phase diagram

For binary systems, when three phases are present, there are no degrees of freedom, since

$$F = 3 - P = 3 - 3 = 0 \qquad (2-13)$$

This means that the compositions of all three phases as well as the temperature are fixed. This condition is met for a eutectic system by the eutectic isotherm; for the Cu-Ag system (Figure 2.3), it is the horizontal line that extends between points B and G. At this temperature, 780 °C, the points at which each of the α, L, and β phase fields touch the isotherm line correspond to the respective phase compositions; namely, the composition of the α phase is fixed at 7.9 *wt.* % Ag, that of the liquid at 71.9 *wt.* % Ag, and that of the β phase at 91.2 *wt.* % Ag. Thus, three-phase equilibrium will not be represented by a phase field, but rather by the unique horizontal isotherm line. Furthermore, all three phases will be in equilibrium for any alloy composition that lies along the length of the eutectic isotherm (*e.g.*, for the Cu-Ag system at 780 °C and compositions between 7.9 and 91.2 *wt.* % Ag).

One use of the Gibbs phase rule is in analyzing for non-equilibrium conditions. For example, a microstructure for a binary alloy that developed over a range of

Chapter 2　Phase Diagrams

temperature and consisting of three phases is a non-equilibrium one; under these circumstances, three phases will exist only at a single temperature.

2.7.6　Iron-Iron Carbide Phase Diagram

A portion of the iron-carbon ($Fe - Fe_3C$) phase diagram is presented in Figure 2.5. Pure iron, upon heating, experiences two changes in crystal structure before it melts. At room temperature the stable form, called ferrite, or an iron, has a BCC crystal structure. Ferrite experiences a polymorphic transformation to FCC austenite, or γ iron, at 912 ℃ (1 674 ℉). This austenite persists to 1 394 ℃ (2 541 ℉), at which temperature the FCC austenite reverts back to a BCC phase known as δ ferrite, which finally melts at 1 538 ℃ (2 800 ℉). All these changes are apparent along the left vertical axis of the phase diagram.

The composition axis in Figure 2.5 extends only to 6.70 *wt*. % C; at this concentration the intermediate compound iron carbide, or cementite (Fe_3C), is formed, which is represented by a vertical line on the phase diagram. Thus, the iron-carbon system may be divided into two parts: an iron-rich portion, as in Figure 2.5; and the other (not shown) for compositions between 6.70 and 100 *wt*. % C (pure graphite). In practice, all steels and cast irons have carbon contents less than 6.70 *wt*. % C; therefore, we consider only the iron-iron carbide system. Figure 2.5

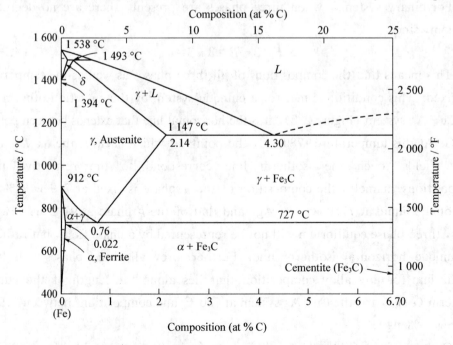

Figure 2.5　The iron-iron carbide phase diagram

would be more appropriately labeled the Fe−Fe$_3$C phase diagram, since F$_3$C is now considered to be a component. Convention and convenience dictate that composition still be expressed in "*wt*. % C" rather than "*wt*. % Fe$_3$C", 6.7 *wt*. % C corresponds to 100 *wt*. % Fe$_3$C.

Carbon is an interstitial impurity in iron and forms a solid solution with each of α and δ ferrites, and also with austenite, as indicated by α, β, and γ single-phase fields in Figure 2.5. In the BCC α ferrite, only small concentrations of carbon are soluble; the maximum solubility is 0.022 *wt*. % at 727 ℃ (1 341 ℉). The limited solubility is explained by the shape and size of the BCC interstitial positions, which make it difficult to accommodate the carbon atoms. Even though present in relatively low concentrations, carbon significantly influences the mechanical properties of ferrite. This particular iron-carbon phase is relatively soft, may be made magnetic at temperatures below 768 ℃ (1 414 ℉), and has a density of 7.88 g · cm^{-3}.

The austenite, or γ phase of iron, when alloyed with just carbon, is not stable below 727 ℃ (1 341 ℉), as indicated in Figure 2.5. The maximum solubility of carbon in austenite, 2.11 *wt*. %, occurs at 1 148 ℃ (2 098 ℉). This solubility is approximately 100 times greater than the maximum for BCC ferrite, since the FCC interstitial positions are larger, and therefore, the strains imposed on the surrounding iron atoms are much lower. As the discussions that follow demonstrate, phase transformations involving austenite are very important in the heat treating of steels. In passing, it should be mentioned that austenite is nonmagnetic.

The δ ferrite is virtually the same as a ferrite, except for the range of temperatures over which each exists. Since the δ ferrite is stable only at relatively high temperatures, it is of no technological importance and is not discussed further.

Cementite (Fe$_3$C) forms when the solubility limit of carbon in α ferrite is exceeded below 727 ℃ (1 341 ℉) (for compositions within the α + Fe$_3$C phase region). As indicated in Figure 2.5, Fe$_3$C will also coexist with the γ phase between 727 and 1 148 ℃ (1 341 and 2 098 ℉). Mechanically, cementite is very hard and brittle; the strength of some steels is greatly enhanced by its presence.

Strictly speaking, cementite is only metastable; that is, it will remain as a compound indefinitely at room temperature. But if heated to between 65 and 700 ℃ (1 200 and 1 300 ℉) for several years, it will gradually change or transform into α iron and carbon, in the form of graphite, which will remain upon subsequent cooling to room temperature. Thus, the phase diagram 2.5 is not a true equilibrium one because cementite is not an equilibrium compound. However, in as much as the decomposition rate of cementite is extremely sluggish, virtually all the carbon in steel will be as Fe_3C instead of graphite, and the iron-iron carbide phase diagram is, for all practical purposes, valid.

The two-phase regions are labeled in Figure 2.5. It may be noted that one eutectic exists for the iron-iron carbide system, at 4.30 $wt.$ % C and 1 148 ℃ (2 098 ℉); for this eutectic reaction,

$$L \xrightleftharpoons[\text{heating}]{\text{cooling}} \gamma + Fe_3C \qquad (2-14)$$

The liquid solidifies to form austenite and cementite phases. Of course, subsequent cooling to room temperature will promote additional phase changes.

It may be noted that a eutectoid invariant point exists at a composition of 0.77 $wt.$ % C and a temperature of 727 ℃ (1 341 ℉). This eutectoid reaction may be represented by

$$\gamma(0.77\ wt.\ \%\ C) \xrightleftharpoons[\text{heating}]{\text{cooling}} \alpha(0.022\ wt.\ \%\ C) + Fe_3C(6.70\ wt.\ \%\ C) \quad (2-15)$$

Or, upon cooling, the solid γ phase is transformed into α iron and cementite. The eutectoid phase changes described by the above are very important, being fundamental to the heat treatment of steels, as explained in subsequent discussions.

Ferrous alloys are those in which iron is the prime component, but carbon as well as other alloying elements may be present. In the classification scheme of ferrous alloys based on carbon content, there are three types: iron, steel, and cast iron. Commercially pure iron contains less than 0.008 $wt.$ % C and, from die phase diagram, is composed almost exclusively of the ferrite phase at room temperature. The iron-carbon alloys that contain between 0.008 $wt.$ % C and 2.11 $wt.$ % C are classified as steels. In most steels the microstructure consists of both α and Fe_3C phases. Upon cooling to room temperature, an alloy within this composition range must pass through at least a portion of the γ phase field; distinctive microstructures are subsequently produced. Although a steel alloy may contain as much as 2.11 $wt.$ % C, in practice, carbon concentrations rarely exceed 1.0 $wt.$ %.

Notes

① iron-carbon system 铁-碳体系
② copper-zinc brass 铜-锌铜
③ sugar-water 糖水
④ heterogeneous systems 异构系统
⑤ copper-silver system 铜-银体系

Vocabulary

abscissa [æbˈsɪsə] $n.$ [数]横坐标;横线

analogy [əˈnælədʒɪ] $n.$ 类比;类推;类似

Chapter 2　Phase Diagrams

austenite ['ɒstəˌnaɪt] n. [材]奥氏体
binary ['baɪnəri] adj. [数]二进制的；二元的，二态的
brass [brɑːs] n. 黄铜；黄铜制品
cementite [sɪ'mentaɪt] n. [材]渗碳体
coexist [ˌkəʊɪɡ'zɪst] vi. 共存，同时存在，并存
compute [kəm'pjuːt] n. 计算；估计；推断 vt. 计算；估算；用计算机计算 vi. 计算；估算；推断
conjunction [kən'dʒʌŋkʃən] n. 结合；同时发生
constitutional [ˌkɒnstɪ'tjuːʃənl] adj. 本质的；体质上的
container [kən'teɪnə] n. 集装箱；容器
correlation [ˌkɒrə'leʃən] n. [数] 相关，关联；相互关系
crystallization [ˌkrɪstəlaɪ'zeɪʃən] n. 晶化，结晶
derivation [ˌderɪ'veɪʃən] n. 推导
diagram ['daɪəɡræm] n. 图表；图解
disparity [dɪ'spærɪti] n. 不同；不一致，不等
dissimilar [dɪ'sɪmɪlə] adj. 不相似的，不同的，不一样的，异样的
distribution [ˌdɪstrɪ'bjuːʃən] n. 分布；分配
entropy ['entrəpi] n. [热]熵（热力学函数）
equilibrium [ˌiːkwɪ'lɪbrɪəm] n. 平衡（状态）；均衡
eutectic [juː'tektɪk] n. 共熔合金 adj. 共熔的；容易溶解的
eutectoid [juː'tektɔɪd] n. [冶]类低共熔体 adj. 共析的；[冶]类低共熔体
fraction ['frækʃən] n. 分数；部分，小部分；稍微

Gibbs [ɡɪbz] n. [物理学]吉布斯
graphical ['ɡræfɪkəl] adj. 图解的；绘画的
graphite ['ɡræfaɪt] n. [矿]石墨 vt. 用石墨涂（或掺入等）
heterogeneous [ˌhetərəʊ'dʒiːniəs] adj. [化学]多相的；异种的；[化学]不均匀的；由不同成分形成的
horizontal [ˌhɒrɪ'zɒntəl] adj. 水平的；地平线的；同一阶层的 n. 水平线，水平面；水平位置
imperceptible [ˌɪmpə'septɪbəl] adj. 感觉不到的；极细微的
intermix [ˌɪntə'mɪks] vt. 使……混杂；使……混合 vi. 混杂；混合
interpretation [ɪnˌtɜːprɪ'teɪʃən] n. 解释；翻译；演出
intersection [ˌɪntə'sekʃən] n. 交叉；十字路口；交集；交叉点
interstitial [ˌɪntə'stɪʃəl] n. 填隙原子；节间 adj. 间质的；空隙的；填隙的
invariant [ɪn'veərɪənt] n. [数]不变量；[计]不变式 adj. 不变的
isomorphous [ˌaɪsəʊ'mɔːfəs] adj. [生物]同形的；[晶体]同晶的
isotherm ['aɪsəʊθɜːm] n. 等温线
liquidus ['lɪkwɪdəs] n. [热]液相线 adj. 液体的，液态的
lowercase [ˌləʊə'keɪs] n. 小写字母；小写字体 adj. 小写字体的 vt. 用小写字体书写
metallurgical [ˌmetə'lɜːdʒɪkl] adj. 冶金的；冶金学的
metastable [ˌmetə'steɪbəl] adj. [物][化学]亚稳的；相对稳定的
molten ['məʊltən] adj. 熔化的；铸造的；炽热的

Chapter 2 Phase Diagrams

multiphase ['mʌltɪfeɪz] n. 多相 adj. [电]多相的;多方面的

nomenclature [nəʊ'meŋklətʃə] n. 命名法;术语

nonmagnetic [ˌnɒnmæg'netɪk] n. 非磁性物 adj. 无磁性的

ordinate ['ɔːdɪnət] n. [数]纵坐标

parameter [pə'ræmɪtə] n. 参数;系数;参量

plot [plɒt] vt. 绘图;划分;标绘

polish ['pɒlɪʃ] v. 磨光;修改;润色 vt. 磨光,使发亮 vi. 擦亮,变光滑

polymorphic [ˌpɒlɪ'mɔːfɪk] adj. [生物]多态的;[生物]多形的;多形态的;[化学]多晶形的(等于 polymorphous)

radii ['reɪdɪaɪ] n. 半径(radius 的复数)

randomness ['rændəmnɪs] n. 随意;无安排;不可测性

saturated ['sætʃəreɪtɪd] v. 使渗透,使饱和(saturate 的过去式) adj. 饱和的;渗透的

silver ['sɪlvə] n. 银

sluggish ['slʌgɪʃ] n. 市况呆滞;市势疲弱 adj. 萧条的;迟钝的;行动迟缓的;懒惰的

solidus ['sɒlɪdəs] n. 斜线

solubility [ˌsɒljʊ'bɪlɪtɪ] n. 溶解度

soluble ['sɒljʊbəl] adj. [化学]可溶的,可溶解的

solute ['sɒljuːt] n. [化学]溶质;[化学]溶解物 adj. 溶解的

solution [sə'luːʃən] n. 溶液;溶解

solvent ['sɒlvənt] n. 溶剂 adj. 有溶解力的

solvus ['sɒlvəs] n. [物化]溶线;固溶体分解曲线

spontaneous [spɒn'teɪnɪəs] adj. 自发的;自然的

stable ['steɪbəl] adj. 稳定的

stipulate ['stɪpjʊleɪt] vi. 规定;保证 vt. 规定;保证

substitutional [ˌsʌbstɪtjuːʃənəl] adj. 取代的;代用的;代理的

subtract [səb'trækt] vt. 减去;扣掉

syrup ['sɪrəp] n. 糖浆,果汁;含药糖浆

terminology [ˌtɜːmɪ'nɒlədʒɪ] n. 术语,术语学

thermodynamic [ˌθɜːməʊdaɪ'næmɪk] adj. 热力学的;使用热动力的

trivial ['trɪvɪəl] adj. [数学]数学上最简单的,平凡的,所有变元都为零的

variables ['verɪəbl] n. [数]变量

Exercises

1. Translate the following Chinese phrases into English

(1) 相图 (2) 机械性能 (3) 钢水
(4) 合金成分 (5) 固溶体 (6) 基质材料(主材料)
(7) 溶解极限 (8) 单相系统 (9) 晶界
(10) 自由能 (11) 空间排列 (12) 铜镍体系
(13) 二元共晶系统 (14) 吉布斯相定律 (15) 相成分

2. Translate the following English phrases into Chinese

(1) phase structure (2) binary phase diagram(s)

(3) solute atom(s) (4) substitutional (or interstitial) position(s)
(5) phase equilibrium (6) metastable state
(7) phase boundary line(s) (8) invariant point
(9) eutectic reaction (10) eutectic isotherm
(11) iron-carbon phase diagram (12) cast iron

3. Translate the following Chinese sentences into English

(1) 如果时间短到不足以形成稳态或者亚稳态相,至少可以用相图来做一些判断,比如,判定首先析出的固相和最后固化的液相。

(2) 有限固溶是指只有一定量的某种物质能完全溶解到另一种物质中。

(3) 阐明了化学势在多组分多相平衡体系及化学反应平衡体系中的应用。

(4) 这两种化合物都不是金属间化合物。

(5) 并发现对于合金凝固,在固相线温度一定的情况下,液相线温度的高低对凝固速度几乎没有影响。

(6) 所以这一项我们可以替代它,用从气相到液相变化的吉布斯自由能。

4. Translate the following English sentences into Chinese

(1) Once the alloy cools past this line, some tin will precipitate to form a new tin—rich phase. This may take the form of fine particles dispersed in the matrix, or in some cases, might be concentrated at the boundaries between grains.

(2) By intentionally introducing solid substitutional atoms into an existing lattice, we can control the properties of the new alloy.

(3) In this diagram the solid solution based on metal C is called the α phase.

(4) There are also many less common ones based on metals with much higher costs and special properties such as beryllium, magnesium, tungsten, uranium, etc.

(5) All eutectics do not grow to produce lamellar structures.

(6) If the carbon content of the steel is known, the proper temperature to which the steel should be heated may be obtained by reference to the iron-iron carbide phase diagram.

5. Translate the following Chinese essay into English

每门科学都有一个独特的词汇与之相关,热力学也不例外。基本概念的精确定义对于发展科学和防止可能造成的误解形成了一个合理的基础。这一章开始我们概述了热力学和单元系统,并继续讨论了一些基本概念,如系统、状态和理想状态,平衡、过程、能量和能量的各种形式,我们也讨论了温度和我呢度范围。

6. Translate the following English essay into Chinese

Josiah Willard Gibbs (1790—1861) was a professor of theology and sacred literature atYale University. He was the father of physical chemist.

In thermodynamics, the Gibbs free energy (Gibbs energy or Gibbs function) is a thermodynamic potential that measures the "useful" or process-initiating work obtainable from an isothermal, isobaric thermodynamic system. The Gibbs free energy is the maximum amount of non-expansion work that can be extracted from a closed system; this maximum can be attained only in a completely reversible process.

Gibbs energy is also the chemical potential that is minimized when a system reaches equilibrium at constant pressure and temperature. As such, it is a convenient criterion of spontaneity for processes with constant pressure and temperature.

扫一扫,查看更多资料

Chapter 3 Solidification and Crystallization

3.1 Introduction

Solidification or crystallization is a process where the atoms are transferred from the disordered liquid state to the more ordered solid state. The rate of the crystallization process is described and controlled by the kinetic laws[①]. These laws give information of the movements of the atoms during the rearrangement. In most cases a driving force[②] is involved that makes it possible to derive the rate of the solidification process.

The aim of this chapter is to study the solidification processes in metals and alloys. The laws of thermodynamics and other fundamental physical laws, which control the solidification, rule these processes.

3.2 Solidification of Metals

In general the solidification of a metal of alloy can be divided into the following steps:

(1) The formation of stable nuclei in the melt (nucleation) (Figure 3.1a).

(2) The growth of nuclei into crystals (Figure 3.1b) and the formation of a grain structure (Figure 3.1c).

The shape that each grain acquires after solidification of the metal depends on many factors, of which thermal gradients are important. The grains are equiaxed since their growth has been about equal in all directions.

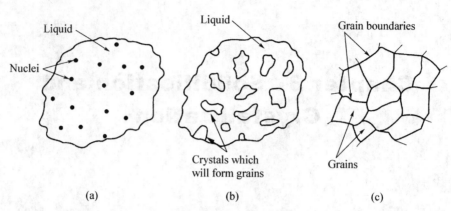

Figure 3.1 Schematic illustration showing the several stages in the solidification of metals: formation of nuclei (a), growth of nuclei into crystals (b) and joining together of crystals to form grains and associated grain boundaries (c). Note that the grains are randomly oriented

3.2.1 Formation of Stable Nuclei in Liquid Metals

The two main mechanisms by which nucleation of solid particles in occur are homogeneous nucleation and heterogeneous nucleation[③].

3.2.2 Homogeneous Nucleation

Homogeneous nucleation is considered first since it is the simplest case of nucleation. Homogeneous nucleation in a liquid melt occurs when the metal itself provides the atoms to form nuclei. Let us consider the case of a pure metal solidifying. When a pure liquid metal is cooled below its equilibrium freezing temperature to a sufficient degree, numerous homogeneous nuclei are created by slow-moving atoms bonding together. Homogeneous nucleation usually requires a considerable amount of undercooling which may be as much as several hundred degrees Celsius for some metals (see Table 3.1). For a nucleus to be stable so that it can grow into a crystal, it must reach a critical size. A cluster of atoms bonded together which is less than the critical size is called an embryo, and one which is larger than the critical size is called a nucleus. Because of their instability, embryos are continuously being formed and redissolved in the molten metal due to the agitation of the atoms.

Chapter 3 Solidification and Crystallization

Table 3.1 Maximum undercooling of some elements

Metal	Melting point / ℃	Maximum observed undercoolingt / ℃
Tin	232	118
Lead	327	60
Aluminum	660	130
Silver	960	227
Copper	1 083	236
Nickel	1 452	319
Iron	1 530	295

In the homogeneous nucleation of a solidifying pure metal, two kinds of energy changes must be considered: (1) the volume (or bulk) free energy released by the liquid to solid transformation, and (2) the surface energy required to form the new solid surfaces of the solidified particles.

When a pure liquid metal such as lead is cooled below its equilibrium freezing temperature, the driving energy for the liquid-to-solid transformation is the difference in the volume (or bulk) free energy ΔG_V of the liquid and that of the solid. If ΔG_V is the change in free energy between the liquid and solid per unit volume of metal, then the free-energy change for a spherical nucleus of radius r is $\frac{4}{3}\pi r^3 \Delta G_V$ since the volume of a sphere is $\frac{4}{3}\pi r^3$. The change in volume free energy vs. radius of an embryo or nucleus is shown schematically in Figure 3.2 as the lower curve and is a negative quantity since energy is released by the liquid to solid transformation.

However, there is an opposing energy to the formation of embryos and nuclei, the energy required to form the surface of these particles. The energy needed to create a surface for these spherical particles ΔG_S, is equal to the specific surface free energy of the particle, γ, times the area of the surface of the sphere, or $4\pi r^2 \gamma$, where $4\pi r^2$ is the surface area of a sphere. This retarding energy for ΔG_S for the formation of the solid particles is shown graphically in Figure 3.2 by an upward curve in the positive upper half of the figure. The total free energy associated with the formation of an embryo or nucleus, which is the sum of the volume free-energy and surface free-energy changes, is shown in Figure 3.2 as the middle curve. In equation form the total free-energy change for the formation of a spherical embryo or nucleus of radius r formed in a freezing pure metal is

$$\Delta G_T = \frac{4}{3}\pi r^3 \Delta G_V + 4\pi r^2 \gamma \qquad (3-1)$$

where ΔG_T: total free energy change; r: radius of embryo or nucleus; ΔG_V:

volume free energy; γ: specific surface free energy.

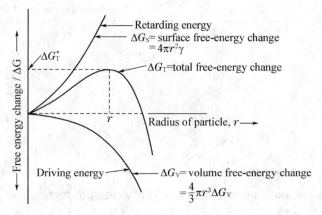

Figure 3.2 Free-energy change ΔG vs. radius of embryo or nucleus created by the solidifying of a pure metal. If the radius of the particle is greater than r^*, a stable nucleus will continue to grow

In nature a system can change spontaneously from a higher to a lower energy state. In the case of the freezing of a pure metal, if the solid particles formed upon freezing have radii less than the critical radius r^*, the energy of the system will be lowered if they redissolve. These small embryos can, therefore, redissolve in the liquid metal. However, if the solid particles have radii greater than r^*, the energy of the system will be lowered when these particles (nuclei) grow into larger particles or crystals (Figure 3.1b). When r reaches the critical radius[④] r^*, ΔG_T has its maximum value of ΔG^* (Figure 3.2).

3.2.3 Critical Radius and Undercooling

The greater the degree of undercooling ΔT below the equilibrium melting temperature of the metal, the greater the change in volume free energy ΔG_V. However, the change in free energy due to the surface energy ΔG_S does not change much with temperature. Thus, the critical nucleus size is determined mainly by ΔG_V. Near the freezing temperature, the critical nucleus size must be infinite since ΔT approaches zero. As the amount of undercooling increases, the critical nucleus size decreases. Figure 3.3 shows the variation in critical nucleus size for copper as a function of undercooling. The maximum amount of undercooling for homogeneous nucleation in the pure metals listed in Table 3.1 is from 60 to 319 ℃.

Chapter 3 Solidification and Crystallization

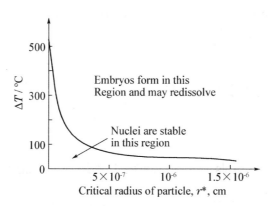

Figure 3.3 Critical radius of copper nuclei vs. degree of undercooling

3.2.4 Heterogeneous Nucleation

Heterogeneous nucleation is nucleation that occurs in a liquid on the surfaces of its container, insoluble impurities, or other structural material which lower the critical free energy required to form a stable nucleus. Since large amounts of undercooling do not occur during industrial casing operations and usually range between 0.1 to 10 ℃, the nucleation must be heterogeneous and not homogeneous.

For heterogeneous nucleation to take place, the solid nucleating agent (impurity solid or container) must be wetted by the liquid metal. Also the liquid should solidify easily on the nucleating agent[5]. Figure 3.4 shows a nucleating agent (substrate) which is wetted by the solidifying liquid creating a low contact angle θ between the solid metal and the nucleating agent. Heterogeneous nucleation takes place on the nucleating agent because the surface energy to form a stable nucleus is lower on this material than if the nucleus is formed in the pure liquid itself (homogeneous nucleation). Since the surface energy is lower for heterogeneous nucleation, the total free-energy change for the formation of a stable nucleus will be lower and the critical size of the nucleus will be smaller. Thus a much smaller

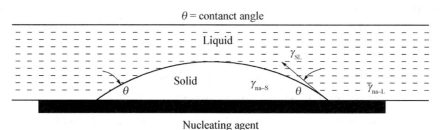

Figure 3.4 Heterogeneous nucleation of a solid on a nucleating agent

(na = nucleating agent, SL = solid-liquid, S = solid, L = liquid, θ = contact angle)

amount of undercooling is required to form a stable nucleus produced by heterogeneous nucleation.

3.3 Growth of Crystals

3.3.1 Growth of Crystals in Liquid Metal and Formation of a Grain Structure

After stable nuclei have been formed in a solidifying metal, these nuclei grow into crystals, as shown in Figure 3.1b. In each solidifying crystal the atoms are arranged in an essentially regular pattern, but the orientation of each crystal varies (Figure 3.1b). When solidification of the metal is finally completed, the crystals join together in different orientations and form crystal boundaries at which changes in orientation take place over a distance of a few atoms (Figure 3.1c). Solidified metal containing many crystals is said to be polycrystalline. The crystals in the solidified metal are called grains, and the surfaces between them, grain boundaries.

The number of nucleation sites available to the freezing metal will affect the grain structure of the solid metal produced. If relatively few nucleation sites are available during solidification, a coarse, or large-grain, structure will be produced. If many nucleation sites are available during solidification, a fine-grain structure will result. Almost all engineering metals and alloys are cast with a fine-grain structure since this is the most desirable type for strength and uniformity of finished metal products.

When a relatively pure metal is cast into a stationary mold without the use of grain refiners, two major of grain structures are usually produced: (1) Equiaxed grains; (2) Columnar grains.

If the nucleation and growth conditions in the liquid metal during solidification are such that the crystals can grow approximately equally in all directions, equiaxed grains will be produced. Equiaxed grains are commonly found adjacent to a cold mold wall, as shown in Figure 3.5. Large amounts of undercooling near the wall create a relatively high concentration of nuclei during solidification, a condition necessary to produce the equiaxed grain structure.

Columnar grains are long, thin coarse grains which are created when a metal solidifies relatively slowly in the presence of a steep temperature gradient. Relatively few nuclei are available when columnar grains are produced. Equiaxed and columnar grains are shown in Figure 3.5. Note that in Figure 3.5b the columnar grains have grown perpendicular to the mold faces since large thermal gradients were present in those directions.

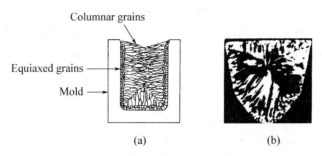

Figure 3.5 Schematic drawing of a solidified metal grain structure produced by using a cold mold (a) and transverse section through an ingot of aluminum alloy 1100 (99.0% Al) cast by the Properzi method (a wheel and belt method) (b). Note the consistency with which columnar grains have grown perpendicular to each mold face

3.3.2 Solidification of Single Crystals

Almost all engineering crystalline materials are composed of many crystals and are therefore polycrystalline. However, there is a few which consist of only one crystal and are therefore single crystals. For example, solid-state electronic components such as transistors and some types of diodes are made from single crystals of semiconducting elements and compounds. Single crystals are necessary for these applications since grain boundaries would disrupt the electrical properties of devices made from semiconducting materials.

In growing single crystals, solidification must take place around a single nucleus so that no other crystals are nucleated and grow. To accomplish this, the interface temperature between the solid and liquid must be slightly lower than the melting point of the solid and liquid temperature must increase beyond the interface. To achieve this temperature gradient, the latent heat of solidification must be conducted through the solidifying solid crystal. The growth rate of the crystal must be slow so that the temperature at the liquid-solid interface is slightly below the melting point of the solidifying solid.

In industry single crystals of silicon 4 to 6 in diameter have been grown for semiconducting device applications. One of the commonly used techniques to produce high quality (minimization of defects) silicon single crystals is the Czochralski method[6]. In this process high-purity polycrystalline silicon is first melted in a noncreative crucible and held at a temperature just above the melting point. A high-quality seed crystal of silicon of the desired orientation is lowered into the melt while it is rotated. Part of the surface of the seed crystal is melted in the liquid to remove the outer strained region and to produce a surface for the liquid to solidify on. The seed crystal continues to rotate and is slowly raised from the melt. As it is raised from the melt, silicon from the liquid in the crucible adheres and grows on the seed crystal, producing a much larger diameter single crystal of silicon (Figure 3.6).

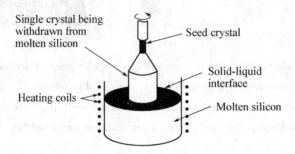

Figure 3.6　Formation of single crystal of silicon by the Czochralski process

3.3.3　Metallic Solid Solutions

Although very few metals are used in the pure or nearly pure state, a few are used in the nearly pure form. For example, high-purity copper of 99.99 percent purity is used for electronic wires because of its very high electrical conductivity. High-purity aluminum (99.99 % Al) (called superpure aluminum) is used for decorative purposes because it can be finished with a very bright metallic surface. However, most engineering metals are combined with other metals or nonmetals to provide increased strength, higher corrosion resistance, or other desired properties.

A metal alloy, or simply an alloy, is a mixture of two or more metals or a metal (metals) and a nonmetal (nonmetals). Alloys can have structures that are relatively simple, such as that of cartridge brass which is essentially a binary alloy (two metals) of 70 % Cu and 30 % Zn. On the other hand, alloys can be extremely complex, such as the nickel-base superalloy Inconel 718[①] used for jet engine parts which has about 10 elements in its nominal composition.

The simplest type of alloy is that of the solid solution. A solid solution is a solid that consists of two or more elements atomically dispersed in a single-phase structure. In general there are two types of solid solutions: substitutional and interstitial.

3.3.3.1 Substitutional Solid Solutions

In substitutional solid solutions[⑧] formed by two elements, solute atoms can substitute for parent solvent atoms in a crystal lattice. Figure 3.7 shows a (111) plane in an FCC crystal lattice in which some solute atoms of one element have substituted for solvent atoms of the parent element. The crystal structure of the parent element or solvent is unchanged, but the lattice may be distorted by the presence of the solute atoms, particularly if there is a significant difference in atomic diameters of the solute and solvent atoms.

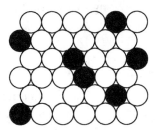

Figure 3.7 Substitutional solid solution
(The dark circles represent one type of atom and the white another.
The plane of atoms is a (1,1,1) plane in an FCC crystal lattice).

The fraction of atoms of one element that can dissolve in another can vary from a fraction of an atomic percent to 100 percent. The following conditions are favorable for extensive solid solubility of one element in another:

(1) The diameters of the atoms of the elements must not differ by more than about 15 percent.

(2) The crystal structures of the two elements must be the same.

(3) There should be no appreciable difference in the electronegativities of the two elements so that compounds will not form.

(4) The two elements should have the same valence.

If the atomic diameters of the two elements which form a solid solution differ, there will be a distortion of the crystal lattice. Since the atomic lattice can only sustain a limited amount of contraction or expansion, there is a limit in the difference in atomic diameters that atoms can have and still maintain a solid solution with the same kind of crystal structure. When the atomic diameters differ by more than about 15 percent, the "size factor" becomes unfavorable for extensive solid solubility.

3.3.3.2 Interstitial Solid Solutions

In interstitial solutions the solute atoms fit into the spaces between the solvent or parent atoms. These spaces or voids are called interstices. Interstitial solid solutions⑨ can form when one atom is much larger than another. Examples of atoms that can form interstitial solid solutions due to their small size are hydrogen, carbon, nitrogen and oxygen.

An important example of an interstitial solid solution is that formed by carbon in FCC γ iron which is stable between 912~1 394 ℃. The atomic radius of γ iron is 0.129 nm and that of carbon is 0.075 nm, and so there is an atomic radius difference of 42 percent. However, in spite of this difference, a maximum of 2.08 percent of the carbon can dissolve interstitially in iron at 1 148 ℃. Figure 3.8a illustrates this schematically by showing distortion around the carbon atoms in the γ iron lattice.

◆ Notes

① kinetic law 动力学定律
② driving force 驱动力
③ heterogeneous nucleation 异相成核
④ critical radius 临界半径
⑤ nucleating agent 成核剂
⑥ Czochralski method 提拉法,切克劳斯基法,单晶提拉法
⑦ Inconel 718 含铌、钼的沉淀硬化型镍铬铁合金
⑧ substitutional solid solution 取代型固溶体
⑨ interstitial solid solution 间隙固溶体
⑩ interstitial void 晶格间隙

◆ Vocabulary

adhere [əd'hɪə] vi. 坚持;依附;粘着 vt. 使粘附

agitation [ˌædʒɪ'teɪʃən] n. 搅拌,搅动

appreciable [ə'priːʃəbl] adj. 可感知的;可评估的;相当可观的

beryllium [bə'rɪlɪəm] n. [化学]铍(符号Be)

cartridge ['kɑːtrɪdʒ] n. 子弹,弹药筒;笔芯,墨盒;暗盒,胶卷盒;录音带盒,唱头

Celsius ['selsɪəs] n. 摄氏度 adj. 摄氏的

columnar [kə'lʌmnə] adj. 柱状的;圆柱的

crucible ['kruːsɪbl] n. 坩埚;熔罐;[冶金学]熔炉底部凹处(容纳金属的地方)

crystallization [ˌkrɪstəlaɪ'zeɪʃən] n. 结晶,结晶,晶化

decorative ['dekərətɪv] adj. 装饰性的;装潢用的

distortion [dɪ'stɔːʃən] n. 变形;[物]失真;扭曲;曲解

equiaxed ['iːkwɪækst] adj. 等轴的,各方等大的,等轴状的

embryo ['embrɪəʊ] n. [胚]胚胎;胚芽;初期 adj. 胚胎的;初期的

gradient ['greɪdɪənt] n. [数][物]梯度;坡度;倾斜度 adj. 倾斜的;步行的

inconel [ˌɪnkə'nel] n. [商标、材料学]因

科镍合金,铬镍铁合金
insoluble [ɪn'sɒljʊbl] adj. 不溶的,不溶解的,不可溶解的
instability [ɪnstə'bɪlɪtɪ] n. 不稳定(性);基础薄弱;不安定
interface ['ɪntəfeɪs] n. 界面;交界面
interstices [ɪn'tɜːstɪsɪz] n. 间隙;缝(interstice 的复数)
interstitial [ɪntə'stɪʃəl] n. 填隙原子;节间 adj. 间质的;空隙的;填隙的
kinetic [kɪ'netɪk] adj. 运动的;动力学的
moderate ['mɒdərət] adj. 稳健的,温和的;适度的,中等的;有节制的
nitrogen ['naɪtrədʒən] n. [化学]氮
nominal ['nɒmɪnəl] adj. 名义上的;有名无实的
nuclei ['njuːklɪaɪ] n. 核心,核子;原子核(nucleus 的复数形式)
polycrystalline
radii ['reɪdɪaɪ] n. 半径(radius 的复数)
redissolve [ˌriːdɪ'zɒlv] vt. 再溶解;再驱散
refiner [rɪ'faɪnə] n. 精炼机;精制者,精炼者
schematic [skiː'mætɪk] n. 原理图;图解视图 adj. 图解的;概要的
substrate ['sʌbstreɪt] n. 基质;基片;底层(等于 substratum)
undercooling [ˌʌndə'kuːlɪŋ] n. [热]过冷;低冷却 vt. 使……过度冷却(undercool 的 ing 形式)

Exercises

1. Translate the following Chinese phrases into English

(1) 凝固与结晶　　(2) 液态金属　　(3) 均相成核

(4) 表面能　　(5) 凝固温度　　(6) 球形颗粒

(7) 纯金属　　(8) 晶体生长　　(9) 晶粒边界

(10) 液固界面　　(11) 镍基高温合金　　(12) 晶格间隙

(13) 柱状晶

2. Translate the following English phrases into Chinese

(1) critical size　　(2) molten metal

(3) volume (or bulk) free energy　　(4) driving energy

(5) equiaxed grain　　(6) high-purity polycrystalline silicon

(7) atomic radius difference　　(8) interstitial solid solutions

(9) nucleating agent　　(10) contact angle

(11) atomic solid solubility　　(12) schematic illustration

3. Translate the following Chinese sentences into English

(1) 晶体自由能是自由原子(离子或者分子)结合成晶体时所放出的能量。

(2) 用经典的成核理论计算了固液表面张力、成核自由能和临界成核半径。

(3) 在固结过程中则将释放出同等的热量。

(4) 晶界对金属的强度和延展性有极大的影响。

(5) 多晶固体材料显微结构的演化与诸多因素密切相关,是一复杂的过程。

4. Translate the following English sentences into Chinese

(1) The results show that pulse electric current can refine solidification macrostructure of the metal, and macrostructure is transformed from bulky columnar grain to fine equiaxed grain.

(2) The nucleation, growth and connection of fatigue cracks were photographed.

(3) Effect of melting temperatures on heterogeneous nucleation of polypropylene nucleated with different nucleating agents was investigated by DSC.

(4) The reason causing bottom-shrinkage is that gravitational segregation results in unusual composition undercooling, the method to eliminate or ease bottom-shrinkage is given.

(5) Otherwise it will be very difficult to redissolve the crystallized out component, since these solutions contain already a high content of solutes.

(6) Both spherical and regular crystalline forms are relatively rare.

5. Translate the following Chinese essay into English

当液体的温度远低于平衡凝固温度以下时,有一个更大的可能性是原子将聚集在一起形成一个比临界半径大的胚胎。此外,在较大的过冷处液体和固体之间有比较大的体积自由能之差。这就减少了原子核的临界尺寸。当过冷变大到足以允许胚胎超过临界尺寸时,均相成核就发生了。

6. Translate the following English essay into Chinese

During solidification, the atomic arrangement changes from at best a short-range order to a long-range order, or crystal structure. Solidification requires two steps: nucleation and growth. Nucleation occurs when a small piece of solid forms from the liquid. The solid must achieve a certain minimum critical size before it is stable. Growth of the solid occurs as atoms from the liquid are attached to the tiny solid until no liquid remains.

扫一扫,查看更多资料

Chapter 4　Phase Transformation

4.1　Introduction

The crystal structure of a metal is not necessarily stable at all temperatures below the melting temperature. This is due to the fact that the solid always assumes the crystal structure with the lowest Gibbs free energy, even if there are other crystal structures with a slightly higher free energy. This holds, in particular, for metals since the binding energy E_0 of a metal depends relatively little on its atomic arrangement. The major contributions to bonding are determined by the electronic structure, and small changes can cause an instability of the crystal structure, for instance by internal fields in ferromagnetic materials. The latter is the cause for the ferromagnetic BCC structure of iron (α - Fe) at low temperatures.

One reason for the versatility of metallic materials lies in the wide range of mechanical properties they possess, which are accessible to management by various means. Three strengthening mechanisms are grain size refinement, solid-solution strengthening and strain hardening. Additional techniques are available wherein the mechanical properties are reliant on the characteristics of the microstructure.

The development of microstructure in both single- and two-phase alloys ordinarily involves some type of phase transformation-an alteration in the number and/or character of the phases. The first portion of this chapter is devoted to a brief discussion of some of the basic principles relating to transformations involving solid phases. In as much as most phase transformations do not occur instantaneously, consideration is given to the dependence of reaction progress on time, or the transformation rate. This is followed by a discussion of the development of two-phase microstructures for iron-carbon alloys. Modified phase diagrams are introduced which permit determination of the microstructure that results from a specific heat treatment. Finally, other microconstituents in addition to pearlite are presented.

Chapter 4 Phase Transformation

4.2 Phase Transformation

A variety of phase transformations are important in the processing of materials, and usually they involve some alteration of the microstructure. For purposes of this discussion, these transformations are divided into three classifications. In the first group are simple diffusion-dependent[①] transformations in which there is no change in either the number or composition of the phases present. These include solidification of pure metal, allotropic transformations recrystallization and grain growth.

In another type of diffusion-dependent transformation, there is some alteration in phase compositions and often in the number of phases present; the final microstructure ordinarily consists of two phases. The eutectoid reaction is of this type; it receives further attention in this chapter.

The third kind of transformation is diffusionless, wherein, a metastable phase is produced. A martensitic transformation which may be induced in some steel alloys, falls into this category.

4.2.1 Kinetics of Solid-State Reaction

Most solid-state transformations do not occur instantaneously because obstacles impede the course of the reaction and make it dependent on time. For example, since most transformations involve the formation of at least one new phase that has a composition and/or crystal structure different from that of the parent one; some atomic rearrangements via diffusion are required. Diffusion is a time-dependent[②] phenomenon. A second impediment to the formation of a new phase is the increase m energy associated with the phase boundaries that are created between parent and product phases.

From a microstructural standpoint, the first process during a phase transformation is nucleation-the formation of very small (often submicroscopic) particles or nuclei of the new phase, which are capable of growing. Favorable positions for the formation of these nuclei are imperfection sites, especially grain boundaries. The second stage is growth, in which the nuclei increase in size; during this process, of course, an amount of the parent phase disappears. The transformation reaches completion if growth of these new phase particles is allowed to proceed until the equilibrium fraction is attained.

Chapter 4 Phase Transformation

As would be expected, the time dependence of the transformation rate (which is often termed the kinetics of a transformation) is an important consideration in the heat treatment of materials. With many kinetic investigations, the fraction of reaction that has occurred is measured as a function of time, while the temperature is maintained constant. Transformation progress is usually ascertained by either microscopic examination or measurement of some physical property (such as electrical conductivity) the magnitude of which is distinctive of the new phase. Data are plotted as the fraction of transformed material versus the logarithm of time; an S-shaped curve shown in Figure 4.1 represents the typical kinetic behavior for most solid-state reactions. Nucleation and growth stages are indicated in the figure.

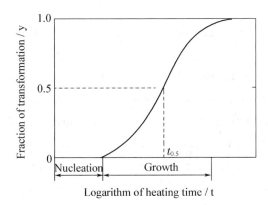

Figure 4.1 **Plot of fraction reacted versus the logarithm of time typical of many solid-state transformations in which temperature is held constant**

For solid-state transformations displaying the kinetic behavior in Figure 4.1, the fraction of transformation y is a function of time t as follows:

$$y = 1 - \exp(-kt^n) \quad (4-1)$$

where k and n are time-independent constants for the particular reaction. The above expression is often referred to as the Avrami equation[3].

By convention, the rate of a transformation r is taken as the reciprocal of time required for the transformation to proceed halfway to completion, $t_{0.5}$, or

$$r = \frac{1}{t_{0.5}} \quad (4-2)$$

This $t_{0.5}$ is also noted in Figure 4.1.

Temperature is one variable in a heat treatment process that is subject to control, and it may have a profound influence on the kinetics and thus on the rate of a transformation. For most reactions and over specific temperature ranges, rate increases with temperature according to

$$r = A e^{-Q/RT} \quad (4-3)$$

where

R = the gas constant.

T = absolute temperature.

A = a temperature-independent[④] constant.

Q = an activation energy for the particular reaction.

It may be recalled that the diffusion coefficient has the same temperature dependence. Processes of which the rates exhibit this relationship with temperature are sometimes termed thermally activated.

4.2.2 Multiphase Transformations

Phase transformations may be wrought in metal alloy systems by varying temperature, composition, and the external pressure; however, temperature changes by means of heat treatments are most conveniently utilized to induce phase transformations. This corresponds to crossing a phase boundary on the composition-temperature phase diagram as an alloy of given composition is heated or cooled.

During a phase transformation, an alloy proceeds toward an equilibrium state that is characterized by the phase diagram in terms of the product phases, their compositions, and relative amounts. Most phase transformations require some finite time to go to completion, and the speed or rate is often important in the relationship between the heat treatment and the development of microstructure. One limitation of phase diagrams is their inability to indicate the time period required for the attainment of equilibrium.

The rate of approach to equilibrium for solid systems is so slow that true equilibrium structures are rarely achieved. Equilibrium conditions are maintained only if heating or cooling is carried out at extremely slow and unpractical rates. For the cooling other than equilibrium, transformations are shifted to lower temperatures than indicated by the phase diagram; for heating, the shift is to higher temperatures. These phenomena are termed supercooling and superheating, respectively. The degree of each depends on the rate of temperature change; the more rapid the cooling or heating, the greater the supercooling or superheating. For example, for normal cooling rates the iron-carbon eutectoid reaction is typically displaced 10 to 20 ℃ (50 to 68 ℉) below the equilibrium transformation temperature.

For many technologically important alloys, the preferred state or microstructure is a metastable one, intermediate between the initial and equilibrium states; on occasion, a structure far removed from the equilibrium one is desired. It thus becomes imperative to investigate the influence of time on phase transformations. This kinetic formation is, in many instances, of greater value than knowledge of the final equilibrium state.

4.3 Microstructural and Property Changes in Iron-Carbon Alloys

Some of the basic kinetic principles of solid-state transformations are now extended and applied specifically to iron-carbon alloys in terms of the relationships between heat treatment, the development of microstructure and mechanical properties. This system has been chosen because it is familiar and because a wide variety of microstructures and mechanical properties are possible for iron-carbon (or steel) alloys.

4.3.1 Isothermal Transformation Diagrams

(1) Pearlite

Consider again the iron-iron carbide eutectoid reaction

$$\gamma(0.77\ wt.\%\ C) \xrightleftharpoons[\text{heating}]{\text{cooling}} \alpha(0.022\ wt.\%\ C) + Fe_3C(6.70\ wt.\%\ C) \quad (4-4)$$

which is fundamental to the development of microstructure in steel alloys. Upon cooling, austenite, having an intermediate carbon concentration, transforms to a ferrite phase, having much lower carbon content, and also to cementite, with a much higher carbon concentration. For carbon atoms to selectively segregates in the cementite phase, diffusion, necessary. Carbon atoms diffuse away from the ferrite regions and to the cementite layers, to give a concentration of $6.70\ wt.\%\ C$, as the pearlite extends from the grain boundary into the unreacted austenite grains. The layered pearlite forms because the carbon atoms need diffuse only minimal distances with this structure.

Temperature plays an important role in the rate of the austenite-to-pearlite transformation. The temperature dependence for an iron-carbon alloy of eutectoid composition is indicated in Figure 4.2, which plots S-shaped curves of the percentage transformation versus the logarithm of time at three different temperatures. For each curve, data is collected after rapidly cooling a specimen composed of 100% austenite to the temperature indicated; that temperature was maintained constant throughout the course of the reaction.

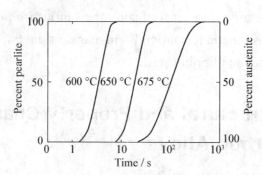

Figure 4.2 For an iron-carbon alloy of eutectoid composition (0.77 *wt.* %) isothermal fraction reacted versus the logarithm of time for the austenite-to-pearlite transformation

A more convenient way of representing both the time and temperature dependence of this transformation is in the bottom portion of Figure 4.3. Here, the vertical and horizontal axes are, respectively, temperature and the logarithm of time. Two solid curves are plotted; one represents the time required at each temperature for the initiation or start of the transformation; the other is for the transformation completion. The dashed curve corresponds to 50 % of transformation completion. These curves were generated from a series of plots of the percentage transformation versus the logarithm of time taken over a range of temperatures. The S-shaped curve [for 675 ℃ (1 247 °F)], in the upper portion of Figure 4.3, illustrates how the data transfer is made.

When interpreting this diagram, note firstly that the eutectoid temperature [727 ℃ (1 341 °F)] is indicated by a horizontal line; at temperatures above the eutectoid and for all time, only austenite will exist, as indicated in the figure. The austenite-to-pearlite transformation will occur only if an alloy is super-cooled to below the eutectoid; as indicated by the curves, the time necessary for the transformation to begin and then end depends on temperature. The start and finish curves are nearly parallel, and they approach the eutectoid line asymptotically. On the left of the transformation start curve only austenite (which is unstable) will be present, whereas on the right of the finish curve, only pearlite will exist. In between, the austenite is in the process of transforming to pearlite, and thus both microconstituents will be present.

According to Equation 4 – 2, the transformation rate at some particular temperature is inversely proportional to the time required for the reaction to proceed to 50 % completion (to the dashed line in Figure 4.3). That is, the shorter this time, the higher the rate. Thus, from Figure 4.3, at temperatures just below the eutectoid (corresponding to just a slight degree of undercooling) very long time (on the order of 10^5 s) are required for the 50 % transformation, and therefore the reaction rate is very slow. The transformation rate increases with decreasing temperature so that at 540 ℃ (1 000 ℉) only about 3 s is required for the reaction to go to 50 % completion.

This rate-temperature behavior is in apparent contradiction of Equation 4 - 3, which stipulates that rate increases with increasing temperature. The reason for this disparity is that over this range of temperatures (*i. e.*, 540 to 727 ℃), the transformation rate is controlled by the rate of pearlite nucleation, and nucleation rate decreases with rising temperature (*i. e.*, less supercooling). This behavior may be explained by Equation 4 - 3, wherein the activation energy Q for nucleation is a function of temperature and increases with increasing temperature. We shall find that at lower temperatures, the austenite decomposition transformation is diffusion-controlled and that the rate behavior is as predicted by Equation 4 - 3, with temperature-independent activation energy for diffusion.

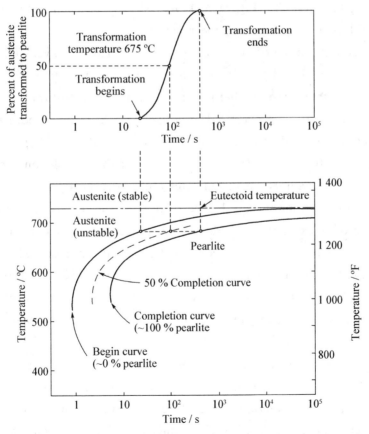

Figure 4.3 Demonstration of how an isothermal transformation diagram (bottom) is generated from percent transformation-versus-logarithm of time measurements (top)

Several constraints are imposed on using diagrams like Figure 4.3. Firstly, this particular plot is valid only for an iron-carbon alloy of eutectoid composition; for other compositions, the curves will have different configurations. In addition, these plots are accurate only for transformations in which the temperature of the alloy is held constant throughout the duration of the reaction. Conditions of constant temperature are termed isothermal; thus, plots such as Figure 4.3 are, referred to as isothermal transformation diagrams, or sometimes as time-temperature-transformation (or T-T-T) plots.

An actual isothermal heat treatment curve (ABCD) is superimposed on the isothermal transformation diagram for a eutectoid iron-carbon alloy in Figure 4.4. Very rapid cooling of austenite to a temperature is indicated by the nearly vertical line AB, and the isothermal treatment at this temperature is represented by the horizontal segment BCD. Of course, time increases from left to right along this line. The transformation of austenite to pearlite begins at the intersection, point C (after approximately 3.5 s), and has reached completion by about 15 s, corresponding to point D. Figure 4.4 also shows schematic microstructures at various time during the progress of the reaction.

The thickness ratio of the ferrite and cementite layers in pearlite is approximately 8 to 1. However, the absolute layer thickness depends on the temperature at which the isothermal transformation is allowed to occur. At temperatures just below the eutectoid, relatively thick layers of both the α-ferrite and Fe_3C phases are produced: this microstructure is called coarse pearlite, and the region at which it forms is indicated on the right of the completion curve in Figure 4.4. At these temperatures, diffusion rates are relatively high. So that during the transformation carbon atoms can diffuse relatively long distances, which results in the formation of thick lamella. With decreasing temperature, the carbon diffusion rate decreases, and the layers become progressively thinner. The thin-layered structure produced in the vicinity of 540 ℃ is termed fine pearlite; this is also indicated in Figure 4.4. For iron-carbon alloys of other compositions, a proeutectoid phase (either ferrite or cementite) will coexist with pearlite.

(2) Bainite

From the discussion of the preceding section, it would seem reasonable to expect the alternating ferrite and cementite lamella to become thinner and thinner as the isothermal transformation temperature is lowered to below that at which fine pearlite forms. Such is not the case: other microconstituents that are products of the austenitic transformation are found to exist at these lower temperatures. One of these microconstituents is called bainite. Furthermore, depending on transformation temperature, two general types of bainite have been observed: upper and lower bainite. Like pearlite, the microstructure of each of these bainites consists of ferrite

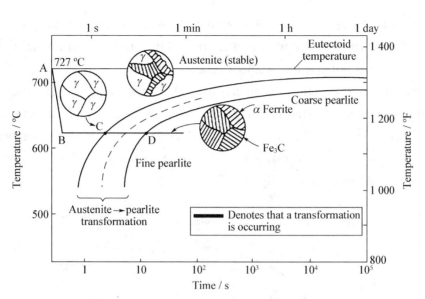

Figure 4.4 Isothermal transformation diagram for a eutectoid iron-carbon alloy, with superimposed isothermal heat treatment curve (ABCD). Microstructure before, during and after the austenite-to-pearlite transformation are shown

and cementite phases; however, their arrangements are different from the alternating lamellar structure found in pearlite.

For temperatures between approximately 300 and 540 ℃, bainite forms as a series of parallel laths (i.e., thin narrow strips) or needles of ferrite that are separated by elongated particles of the cementite phase. Such is termed upper bainite, and its microstructural details are so fine that their resolution is possible only using electron microscopy. At lower temperatures between about 200 and 300 ℃ lower bainite is the transformation product. For lower bainite, the ferrite phase exists as thin plates (instead of laths as with upper bainite), and narrow cementite particles (as very fine rods or blades) form within these ferrite plates.

The time-temperature dependence of the bainite transformation may also be represented on the isothermal transformation diagram. It occurs at temperatures below those at which pearlite forms; begin-, end- and half-reaction curves are just extensions of those for the pearlitic transformation, as shown in Figure 4.5, the isothermal transformation diagram for an iron-carbon alloy of eutectoid composition that has been extended to lower temperatures. All three curves are C-shaped and have a "nose" at point N, where the rate of transformation is a maximum. As may be noted, whereas pearlite forms above the nose, that is, over the temperature range of about 540 to 727 ℃ (1 000 to 1 341°F)—for isothermal treatments at temperatures between about 215 and 540 ℃ (420 and 1 000 °F), bainite is the transformation product. It should also be noted that pearlitic and bainitic transformations are really competitive with each other, and once some portion of an alloy has transformed to either pearlite or bainite, transformation to the other microconstituent is not possible without reheating to form austenite.

In passing, it should be mentioned that the kinetics of the bainite transformation (below the nose in Figure 4.5) obey Equation 4-3; that is, rate ($1/t_{0.5}$, Equation 4-3) increases exponentially with rising temperature. Furthermore, the kinetics of many solid-state transformations is represented by this characteristic C-shaped curve (Figure 4.5).

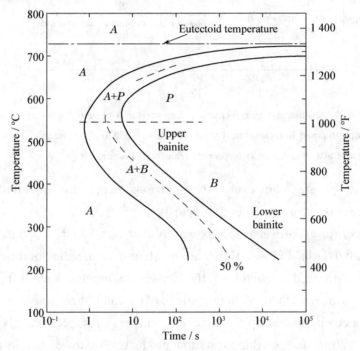

Figure 4.5 Isothermal transformation diagram for an iron-carbon alloy of eutectoid composition, including austenite-to-pearlite (A—P) and austenite-to-bainite (A—B) transformations

(3) Martensite

Yet another microconstituent or phase called martensite is formed when austenitized iron-carbon alloys are rapidly cooled (or quenched) to a relatively low temperature (in the vicinity of the ambient). Martensite is a non-equilibrium single-phase structure that results from a diffusionless transformation of austenite. It may be thought of as a transformation product that is competitive with pearlite and bainite. The martensitic transformation occurs when the quenching rate is rapid enough to prevent carbon diffusion. Any diffusion whatsoever will result in the formation of ferrite and cementite phases.

The martensitic transformation is not well understood. However, large numbers of atoms experience cooperative movements, in that there is only a slight displacement of each atom relative to its neighbors that the FCC austenite experiences a polymorphic transformation to a body-centered tetragonal (BCT)[5] martensite. A unit cell of this crystal structure is simply a body-centered cube that has been elongated along one of its dimensions; this structure is distinctly different from that for BCC ferrite. All the carbon atoms remain as interstitial impurities in martensite; as much, they constitute a supersaturated solid solution that is capable of rapidly transforming to other structures if heated to temperatures at which diffusion rates become appreciable. Many steels, however, retain their martensitic structure almost indefinitely at room temperature.

The martensitic transformation is not, however, unique to iron-carbon alloys. It is found in other systems and is characterized, in part, by the diffusionless transformation. Since the martensitic transformation does not involve diffusion, it occurs almost instantaneously; the martensite grains nucleate and grow at a very rapid rate-the velocity of sound within the austenite matrix. Thus the martensitic transformation rate, for all practical purposes, is time independent.

Two distinctly different martensitic microstructures are found in iron-carbon alloys: lath and lenticular. For alloys containing less than about 0.6 wt. % C, the martensite grains form as laths (i.e., long and thin plates, like blades of grass) that form side by side, and are aligned parallel to one another. Furthermore, these laths are grouped into larger structural entities, called blocks; the morphology of this lath martensite is represented schematically in Figure 4.6. Microstructural details of this type of martensite are too fine to be revealed by optical microscopy, and, therefore, electron micrographic techniques must be employed.

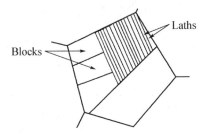

Figure 4.6 Schematic diagram showing the microstructural features of lath or massive martensite

Lenticular (or plate) martensite is typically found in iron-carbon alloys containing greater than approximately 0.6 wt. % C. With this structure the martensite grains take on a needlelike (i.e., lenticular) or platelike appearance. Here the lenticular martensite grains are the dark regions, whereas the white phase is retained austenite that did not transform during the rapid quench.

It should be noted that, as has already been mentioned, both of these types of martensite as well as other microconstituents (e. g., pearlite and bainite) can coexist.

Being a non-equilibrium phase, martensite does not appear on the iron-iron carbide phase diagram Figure 4.6). The austenite-to-martensite transformation is, however, represented on the isothermal transformation diagram. Since the martensitic transformation is diffusionless and instantaneous, it is not depicted in this diagram like the pearlitic and bainitic reactions. The beginning of this transformation is represented by a horizontal line designated M (start) (Figure 4.7). Two other horizontal and dashed lines, labeled M (50 %) and M (90 %), indicate percentages of the austenite-to-martensite transformation. The temperatures at which these lines are located vary with alloy composition, nevertheless, must be relatively low because carbon diffusion must be virtually nonexistent. The horizontal and linear character of these lines indicates that the martensitic transformation is independent of time; it is a function only of the temperature to which the alloy is quenched or rapidly cooled. A transformation of this type is termed an athermal transformation.

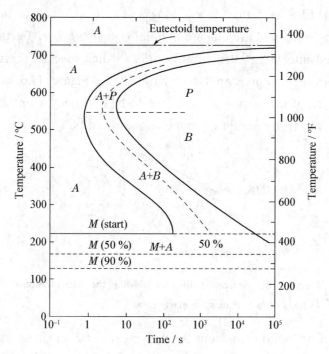

Figure 4.7 The complete isothermal transformation diagram for an iron-carbon alloy of eutectoid composition: A—austenite; B—bainite; M—martensite; P—pearlite

An alloy of eutectoid composition can be taken as an example. This alloy is very rapidly cooled from a temperature above 727 (1 341) to 165 ℃ (330 °F). From the isothermal transformation diagram (Figure 4.7) it may be noted that 50 % of the austenite will immediately transform to martensite at 165 ℃. And as long as the temperature is maintained at this point, there will be no further transformation.

The presence of alloying elements other than carbon (e.g., Cr, Ni, Mo and W) may cause significant changes in the positions and shapes of the curves in the isothermal transformation diagrams. These include (1) shifting, to longer time the nose of the austenite-to-pearlite transformation (and also a proeutectoid phase nose, if such exists), and (2) the formation of a separate bainite nose. These alterations may be observed by comparing Figures 4.7 and 4.8, which are isothermal transformation diagrams for carbon and alloy steels, respectively.

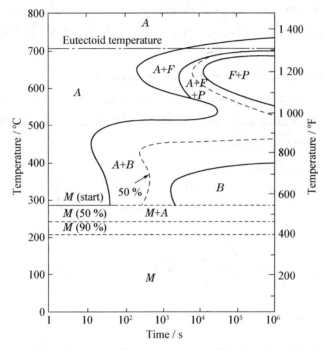

Figure 4.8 Isothermal transformation diagram for a eutectoid iron-carbon alloy steel: A—austenite; B—bainite; M—martensite; P—pearlite

4.3.2 Continuous Cooling Transformation Diagram

Isothermal heat treatments are not the most practical things to conduct because an alloy must be rapidly cooled to and maintained at an elevated temperature from a higher temperature above the eutectoid. Most heat treatments for steels involve the continuous cooling of a specimen to room temperature. An isothermal transformation diagram is valid only for conditions of constant temperature, while diagram must be modified for transformations that occur as the temperature is constantly changing. For continuous cooling, the time required for a reaction to begin and end is delayed. Thus, the isothermal curves are shifted to longer times and lower temperatures, as indicated in Figure 4.9 for an iron-carbon alloy of eutectoid composition. A plot containing such modified beginning and ending reaction curves is termed a continuous cooling transformation (CCT)[⑥] diagram. Some control may be maintained over the rate of temperature change depending on the cooling environment. Two cooling curves corresponding to moderately fast and slow rates are superimposed and labeled in Figure 4.10, again for eutectoid steel. The transformation starts after a time period corresponding to the intersection of the cooling curve with the beginning reaction curve and concludes upon crossing the completion transformation curve. The microstructural products for the moderately rapid and slow cooling rate curves in Figure 4.10 are fine and coarse pearlite, respectively.

Normally, bainite will not form when an alloy of eutectoid composition or, for that matter, any plain carbon steel is continuously cooled to room temperature. This is because all the austenite will have transformed to pearlite by the time the bainite transformation has become possible.

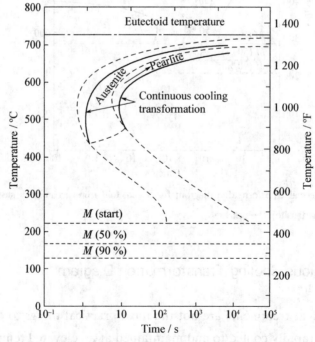

Figure 4.9 Superimposition of isothermal and continuous cooling transformation diagrams for a eutectoid iron-carbon alloy

Thus, the region representing the austenite-pearlite transformation terminates just below the nose (Figure 4.10) as indicated by the curve AB. For any cooling curve passing through *AB* in Figure 4.10, the transformation ceases at the point of intersection; with continued cooling, the unreacted austenite begins transforming to martensite upon crossing the *M* (start) line.

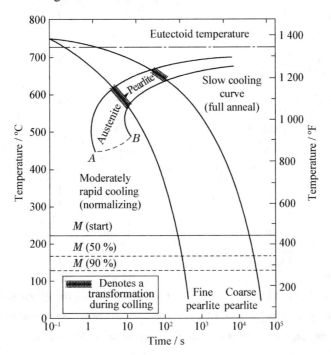

Figure 4.10 Moderately rapid and slow cooling curves superimposed on a continuous transformation diagram for a eutectoid iron-carbon alloy

With regard to the representation of the martensitic transformation, the *M* (start), *M* (50 %), and *M* (90 %) lines occur at identical temperatures for both isothermal and continuous cooling transformation diagrams. This may be verified for an iron-carbon alloy of eutectoid composition by comparison of Figures 4.7 and 4.9.

For the continuous cooling of a steel alloy, there exists a critical quenching rate, which represents the minimum rate of quenching that will produce a totally martensitic structure. This critical cooling rate, when included on the continuous transformation diagram, will just miss the nose at which the pearlite transformation begins, as illustrated in Figure 4.11. As the figure also shows, only martensite will exist when quenching rate is greater than the critical; in addition, there will be a range of rates over which both pearlite and martensite are produced. Finally, a totally pearlitic structure develops for low cooling rates.

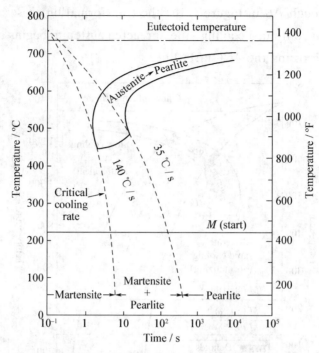

Figure 4.11 **Continuous cooling transformation diagram for a eutectoid iron-carbon alloy and superimposed cooling curves, demonstrating the dependence of the final microstructure on the transformations that occur during cooling**

Carbon and other alloying elements also shift the pearlite (as well as the proeutectoid phase) and bainite noses to longer time, thus decreasing the critical cooling rate. In fact, one of the reasons for alloying steels is to facilitate the formation of martensite so that totally martensitic structures can develop in relatively thick cross sections.

Interestingly enough, the critical cooling rate is diminished even by the presence of carbon. In fact, iron-carbon alloys containing less than about 0.25 *wt*. % carbon are not normally heat treated to form martensite because quenching rates too rapid to be practical are required. Other alloying elements that are particularly effective in rendering steels heat treatable are chromium, nickel, molybdenum, manganese, silicon and tungsten, however, these elements must be in solid solution with the austenite at the time of quenching.

In summary, isothermal and continuous transformation diagrams are m a sense phase diagrams in which the parameter of time is introduced. Each is experimentally determined for an alloy of specified composition, the variables being temperature and time. These diagrams allow prediction of the microstructure after some time period for constant temperature and continuous cooling heat treatments, respectively.

Chapter 4 Phase Transformation

Notes

① diffusion-dependent 与扩散有关的
② time-dependent 与时间有关的
③ Avrami equation 阿夫拉米方程
④ temperature-independent 与温度无关的
⑤ body-centered tetragonal（简称BCT）体心正方
⑥ continuous cooling transformation（简称CCT）连续冷却转变

Vocabulary

allotropic [ˌæləʊˈtrɒpɪk] adj. 同素异形的
ambient [ˈæmbɪənt] n. 周围环境 adj. 周围的;外界的;环绕的
athermal [əˈðɜːml] adj. 不热的;无热的
chromium [ˈkrəʊmɪəm] n. [化]铬
disparity [dɪˈspærɪtɪ] n. 不同;不一致;不等
impede [ɪmˈpiːd] vt. 阻碍;妨碍;阻止
intersection [ˌɪntəˈsekʃən] n. 交叉;十字路口;交集;交叉点
lath [lɑːθ] n. 板条 vt. 给……钉板条
lenticular [lenˈtɪkjʊlə] adj. 透镜的;晶状体的;两面凸的
logarithm [ˈlɒgərɪðəm] n. [数]对数
martensitic [ˌmɑːtɪnˈzɪtɪk] adj. 马氏体的
microconstituents [ˌmaɪkrəʊkɒnˈstɪtjʊənt] n. 显微组分;显微组织成份
molybdenum [məˈlɪbdənəm] n. [化学]钼
nonexistent [ˌnɒnɪgˈzɪstənt] adj. 不存在的
proeutectoid [pruːˈtektwɑːd] n. [冶]先共析体;前共析
quench [kwentʃ] vt. [冶金学]淬火;使骤冷
reciprocal [rɪˈsɪprəkəl] n. [数]倒数;互相起作用的事物 adj. 互惠的;相互的;倒数的,彼此相反的
recrystallization [riːˌkrɪstəlaɪˈzeɪʃən] n. 再结晶
refinement [rɪˈfaɪnmənt] n. [冶]提纯
submicroscopic [ˌsʌbmaɪkrəˈskɒpɪk] adj. 亚微观的
supercooling [ˌsjuːpəˈkuːlɪŋ] n. [热]过冷 v. 使过冷（supercool 的现在分词）
tetragonal [tɪˈtrægənəl] adj. [数]四角形的
unreacted [ˈʌnrɪˈæktɪd] adj. 未反应的

Exercises

1. Translate the following Chinese phrases into English

(1) 相变   (2) 动力学行为   (3) 绝对温度
(4) 扩散系数 (5) 热处理   (6) 相对量

(7) 过冷和过热　　　(8) 平衡转变温度　　　(9) 粗的珠光体
(10) 贝氏体转变　　　(11) 快速淬火　　　　(12) 临界冷却率

2. Translate the following English phrases into Chinese

(1) ferromagnetic materials　　(2) solid-state reaction
(3) S-shaped curve　　(4) temperature-independent constant
(5) iron-carbon eutectoid reaction　　(6) unreacted austenite grain(s)
(7) martensitic structure　　(8) continuous cooling transformation diagram

3. Translate the following Chinese sentences into English

(1) 非共晶固溶体有宽的固化温度范围,且固化过程中晶体的成分在发生变化,所以在模壁上形成细晶层,但是离模壁远的地方,晶粒以枝状形式增长,形状与分叉的树很相像。

(2) 这种层状微结构即薄层状微结构,每层的厚度大约在 1μm。

(3) 这种结晶的固化过程与亚共晶合金类似,不同的是它在降温过程中会穿过"β 相-液态"共存区而不是"α 相-液相"共存区。

(4) 温度一低,结晶生长速度就快起来。

(5) 通常用字母 A 表示奥氏体、字母 F 表示铁素体和字母 G 表示石墨。

(6) 钢和铸铁都有共析、亚共析和过共析这三种组织。

(7) 为了控制奥氏体的转变特性,可以采用合金化和热处理工艺。

(8) 在一定的化学成分和冷却速度下,合金液过冷度的大小受到合金液中存在的晶核的强烈影响。

4. Translate the following English sentences into Chinese

(1) Plain carbon steels are used in low cost, high strength applications where weight and corrosion are not a problem.

(2) An alloy with a composition that lies to the left of the eutectic point on the phase diagram is called a hypoeutectic alloy, and an alloy with a composition that lies to the right of the eutectic point is called hypoeutectic alloy.

(3) The smaller carbides will easily decompose and go into solution in the austenite and produce a more homogeneous austenite structure.

(4) Oxidation will make iron and steel rusty.

(5) The redistribution of carbon atoms by diffusion within the austenite is necessary to allow ferrite which contains very little carbon or cementitie which contains 6.7% carbon to form.

(6) The lower the temperature of bainite formation the finer are these carbides and the structures produced become similar to that of tempered martensite.

5. Translate the following Chinese essay into English

当镍钛诺合金加热到高温时,经热加工成型成所需形状。淬火后它具有马氏体结构。在这种状态下,这种金属并不是一种硬质材料,而是非常柔软的。如果此时将它冷却成型为一种新的形状,它仍会记得其初始形状。将这种柔软的金属加热到其特有的转变温度时,它会突然重新成型为初始的外形。

6. Translate the following English essay into Chinese

Solid-state phase transformations are important means for the adjustment of the microstructure and thus the tuning of the properties of materials. To exploit this tool to full extent, much effort is spent on the modelling of phase transformations. The required models should not be in particular of atomistic nature, but pertain to larger, mesoscopic and even macroscopic scales. However, atomistic simulations can be very useful for the interpretation of the values obtained for the kinetic parameters.

扫一扫,查看更多资料

Part II
Foundation of Material Structures

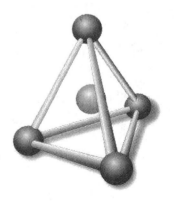

Chapter 5　Crystal Structure

5.1　Introduction

The various types of atomic bonding are determined by the electron structure of the individual atoms. The present discussion is devoted to the level of the structure of materials, specifically, to some of the arrangements that may be assumed by atoms in the solid state. Within this framework, concepts of crystallinity and noncrystallinity are introduced. For crystalline solids the notion of crystal structure is presented, specified in terms of a unit cell①. Crystal structures found in both metals and ceramics are then detailed, along with the scheme by which crystallographic directions and planes are expressed.

5.2　Fundamental Concepts

5.2.1　Space Lattice and Unit Cells

The physical structure of solid materials of engineering importance depends mainly on the arrangements of the atoms, ions or molecules which make up the solid and the bonding forces between them. If the atoms or ions of a solid are arranged in a pattern that repeats itself in three dimensions, they form a solid which is said to have a crystal structure and is referred to as a crystalline solid or crystalline material. Examples of crystalline materials are metals, alloys and some ceramic materials.

Atomic arrangements in crystalline solids can be described by referring the atoms to the points of intersection of a network of lines in three dimensions. Such a network is called a space lattice② (Figure 5.1a) and can be described as an infinite three-dimensional array of points. Each point in the space lattice has identical surroundings. In an ideal crystal the grouping of lattice points about any given point are identical with the grouping about any other lattice point in the crystal lattice. Each space lattice can thus be described by specifying the atom positions in a repeating unit cell, such as the one heavily outlined in Figure 5.1a. The size and shape of the unit cell can be described by three lattice vectors a, b and c, originating from one corner of the unit cell (Figure 5.1b). The axial lengths a, b and c and the inter-axial angles α, β and γ are the lattice constants of the unit cell.

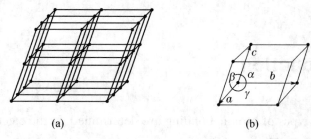

(a) (b)

Figure 5.1 Space lattice of ideal crystalline solid (a) and unit cell showing lattice constants (b)

5.2.2 Crystal Systems and Bravais Lattice[③]

By assigning specific values-for axial lengths and interaxial angles, unit cells of different types can be constructed. Crystallographers have shown that only seven different types of unit cells are necessary to create all point lattices. These crystal systems are listed in Table 5.1.

Table 5.1 Classification of space lattices by crystal system

Crystal system	Axial lengths and inter-axial angles	Space lattice
Cubic	Three equal axes at right angles $a = b = c$, $\alpha = \beta = \gamma = 90°$	Simple cubic, Body-centered cubic Face-centered cubic
Tetragonal	Three axes at right angles, two equal $a = b \neq c$, $\alpha = \beta = \gamma = 90°$	Simple tetragonal Body-centered tetragonal
Orthorhombic	Three unequal axes at right angles $a \neq b \neq c$, $\alpha = \beta = \gamma = 90°$	Simple orthorhombic Body-centered orthorhombic Base-centered orthorhombic Face-centered orthorhombic
Rhombohedral	Three equal axes, equally inclined $a = b = c$, $\alpha = \beta = \gamma \neq 90°$	Simple rhombohedral
Hexagonal	Two equal axes at 120o, third axis at right angles $a = b \neq c$, $\alpha = \beta = 90°$, $\gamma \neq 120°$	Simple hexagonal

Crystal system	Axial lengths and inter-axial angles	(to be continued) Space lattice
Monoclinic	Three unequal axes, one pair not at right angles $a \neq b \neq c$, $\alpha = \gamma = 90° \neq \beta$	Simple monoclinic Base-centered monoclinic
Triclinic	Three unequal axes, unequally inclined and none at right angles $a \neq b \neq c$, $\alpha \neq \beta \neq \gamma \neq 90°$	Simple triclinic

Many of the seven crystal systems have variations of the basic unit cell. A. J. Bravais showed that 14 standard unit cells could describe all possible lattice networks. These Bravais lattices are illustrated in Figure 5.2. There are four basic types of unit cells: simple, body-centered, face-centered and base-centered[④].

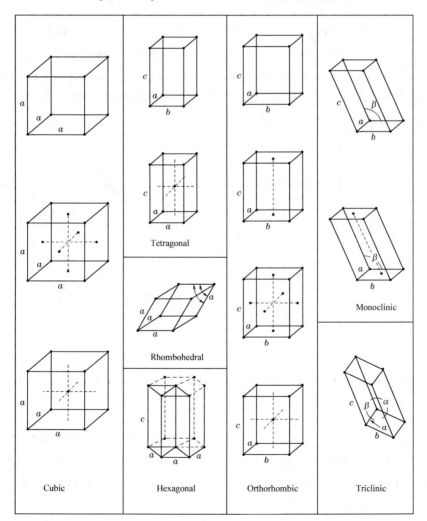

Figure 5.2 The 14 Bravais conventional unit cells grouped according to crystal system. The dots indicate lattice points that, when located on faces or at corners, are shared by other identical lattice unit cells

Chapter 5 Crystal Structure

In the cubic system there are three types of unit cells: simple cubic, body-centered cubic and face-centered cubic. In the orthorhombic system all four types are represented. In the tetragonal system there are only two: simple and body-centered. The face-centered tetragonal unit cell appears to be missing but can be constructed from four body-centered tetragonal unit cells. The monoclinic system has simple and base-centered unit cells, and the rhombohedral, hexagonal, and triclinic systems have only one simple type of unit cell.

5.2.3 Crystallographic Directions and Miller Indies

5.2.3.1 Atom Positions in Unit Cells

To locate atom positions in cubic unit cells, we use rectangular x, y, and z axes. In crystallography the positive x-axis is usually the direction coming out of the paper, the positive y-axis is the direction to the right of the paper, and the positive z-axis is the direction to the top (Figure 5.3). Negative directions are opposite to those just described.

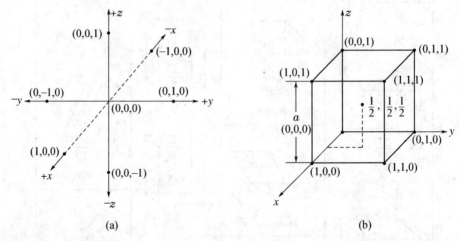

Figure 5.3 Rectangular x, y, and z axes for locating atom positions in cubic unit cells (a). Atom positions in a BCC unit cell (b)

Atom positions in unit cells are located by using unit distances along the x, y and z axes, as indicated in Figure 5.3a. For example, the position coordinates for the atoms in the BCC unit cell are shown in Figure 5.3b. The atom positions for the eight corner atoms of the BCC unit cell are

$$(0, 0, 0) \quad (1, 0, 0) \quad (0, 1, 0) \quad (0, 0, 1)$$
$$(0, 0, 0) \quad (1, 1, 1) \quad (1, 0, 1) \quad (0, 1, 1)$$

The center atom in the BCC unit cell has the position coordinates $\left(\frac{1}{2}, \frac{1}{2}, \frac{1}{2}\right)$. For simplicity sometimes only two atom positions in the BCC unit cell are specified which are (0, 0, 0) and $\left(\frac{1}{2}, \frac{1}{2}, \frac{1}{2}\right)$. The remaining atom positions of the BCC unit cell are assumed to be understood. In the same way the atom positions in the FCC unit cell can be located.

5.2.3.2 Directions in Cubic Unit Cells

Frequently it is necessary to refer to specific directions in crystal lattices. This is especially important for metals and alloys that have properties which vary with crystallographic orientation. For cubic crystals the crystallographic direction indices are the vector components of the direction resolved along each of the coordinate axes and reduced to the smallest integers.

To diagrammatically indicate a direction in a cubic unit cell, we draw a direction vector from an origin which is usually a corner of the cubic cell until it emerges from the cube surface (Figure 5.4). The position coordinates of the unit cell where the direction vector emerges from the cube surface after being converted to integers are the direction indices. The direction indices are enclosed by square brackets with no separating commas.

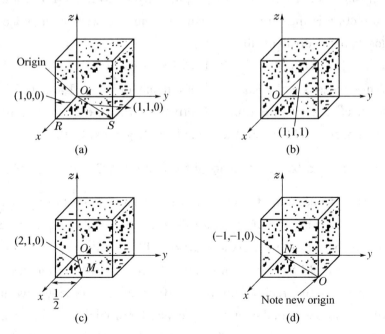

Figure 5.4 Some directions in cubic unit cells

For example, the position coordinates of the direction vector *OR* in Figure 5.4a where it emerges from the cube surface are (1, 0, 0), and so the direction indices for the direction vector *OR* are [100]. The position coordinates of the direction vector *OS* (Figure 5.4a) are (1, 1, 0), and so the direction indices for *OS* are [110]. The position coordinates for the direction vector *OT* (Figure 5.4b) are (1, 1, 1), and so the direction indices of *OT* are [111].

The position coordinates of the direction vector *OM* (Figure 3.4c) are $\left(1, \frac{1}{2}, 0\right)$, and since the direction vectors must be integers, these position coordinates must be multiplied by 2 to obtain integers. Thus, the direction indices of *OM* become $2\left(1, \frac{1}{2}, 0\right) = [210]$. The position coordinates of the vector *ON* (Figure 5.4d) are (-1, -1, 0). A negative direction index is written with a bar over the index. Thus, the direction indices for the vector *ON* are $[\bar{1},\bar{1},0]$. Note that to draw the direction *ON* inside the cube the origin of the direction vector had to be moved to the front lower-right corner of the unit cube (Figure 5.4d).

The letters u, v and w are used in a general sense for the direction indices in the x, y and z directions, respectively, and are written as [u v w]. It is also important to note that all parallel direction vectors have same direction indices.

Directions are said to be crystallographic ally equivalent if the atom spacing along each direction is the same. For example, the following cubic edge directions are crystallographic equivalent directions:

[100], [010], [001], [0 $\bar{1}$ 0], [00 $\bar{1}$], [$\bar{1}$00] = <100>

Equivalent directions are called indices of a family or form. The notation <100> is used to indicate cubic edge directions collectively. Other directions of a form are the cubic body diagonals <111> and the cubic face diagonals <110>.

5.2.3.3 Miller Indices for Crystallographic Planes in Cubic Unit Cells

Sometimes it is necessary to refer to specific lattice planes of atoms within a crystal structure, or it may be of interest to know the crystallographic orientation of a plane or group of planes in a crystal lattice. To identify crystal planes in cubic crystal structures, the Miller notation system is used. The Miller indices of a crystal plane are defined as the reciprocals of the fractional intercepts (with fractions cleared), which the plane makes with the crystallographic x, y and z axes of the three nonparallel edges of the cubic unit cell. The cube edges of the unit cell represent unit lengths, and the intercepts of the lattice planes are measured in terms of these unit lengths.

Chapter 5 Crystal Structure

The procedure for determining the Miller indices for a cubic crystal plane is as follows:

(1) Choose a plane that does not pass through the origin at (0, 0, 0).

(2) Determine the intercepts of the plane in terms of the crystallographic x, y and z axes for a unit cube. These intercepts may be fractions.

(3) Form the reciprocals of these intercepts.

(4) Clear fractions and determine the smallest set of whole number which are in the same ratio as the intercepts. These whole numbers are the Miller indices of the crystallographic plane and are enclosed in parentheses without the use of commas. The notation $(h\ k\ l)$ is used to indicate Miller indices in a general sense, where h, k and l are the Miller indices of a cubic crystal plane for the x, y and z axes, respectively.

Figure 5.5 shows three of the most important crystallographic planes of cubic crystal structures. Let us first consider the shaded crystal plane in Figure 5.5a, which has the intercepts 1, ∞ and ∞ for the x, y and z axes, respectively. We take the reciprocals of these intercepts to obtain the Miller indices, which are therefore 1, 0, 0. Since these numbers do not involve fractions, the Miller indices for this plane are [100], which is read as the one-zero-zero plane. Next let us consider the second plane shown in Figure 5.5b. The intercepts of this plane are 1, 1, ∞. Since the reciprocals of these numbers are 1, 1, 0, which do not involve fractions, the Miller indices of this plane are [1, 1, 0]. Finally, the third plane (Figure 5.5c) has the intercepts 1, 1, 1, which give the Miller indices [111] for this plane.

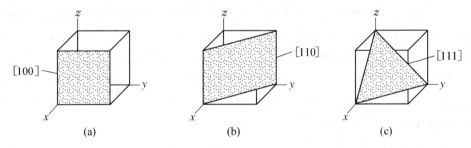

Figure 5.5 Miller indices of some important cubic crystal planes: (a)-[100], (b)-[110] and (c)-[111]

Consider now the cubic crystal plane shown in Figure 5.6 which has the intercepts $\frac{1}{3}$, $\frac{2}{3}$, 1. The reciprocals of these intercepts are 3, $\frac{3}{2}$, 1. Since fractional intercepts are not allowed, these fractional intercepts must be multiplied by 2 to clear the $\frac{3}{2}$ fraction. Thus, the reciprocal intercepts become 6, 3, 2 and the Miller indices are [632].

Chapter 5 Crystal Structure

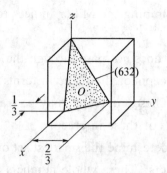

Figure 5.6 Cubic crystal plane (632) which has fractional intercepts

If the crystal plane being considered passes through the origin so that one or more intercepts are zero, the plane must be moved to an equivalent position in the same unit cell and the plane must remain parallel to the original plane. This is possible because all equispaced parallel planes are indicated by the same Miller indices.

If sets of equivalent lattice planes are related by the symmetry of the crystal system, they are called planes of a family or form, and the indices of one plane of the family are enclosed in braces as {hkl} to represent the indices of a family of symmetrical planes. For example, the Miller indices of the cubic surface planes [100], [010] and 001 are designated collectively as a family or form by the notation {100}.

In cubic crystal structures the interplanar spacing between two closed parallel planes with the same Miller indices is designated d_{hkl}, where h, k and l are the Miller indices of the planes. This spacing represents the distance from a selected origin containing one plane and another parallel plane with the same indices which is closest to it. For example, the distance between (110) plane 1 and 2, d_{110}, in Figure 5.7 is AB. Also, the distance between (110) plane 2 and 3 are d_{110} and is length BC in Figure 5.7. From simple geometry, it can be shown that for cubic crystal structures

$$d_{hkl} = \frac{a}{\sqrt{h^2 + k^2 + l^2}} \qquad (5-1)$$

Where d_{hkl}: interplanar spacing between parallel closest planes with Miller indices h, k, and l.

a: lattice constant (edge of unit cube).

h, k, l: Miller indices of cubic planes being considered.

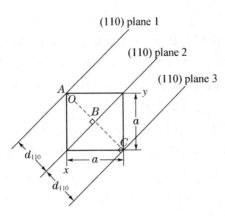

Figure 5.7 Top view of cubic unit cell showing the distance between [110] crystal planes, d_{110}

5.2.3.4 Crystallographic Planes and Directions in Hexagonal Unit Cells

(1) Indices for crystal planes in HCP unit cells

Crystal planes in HCP unit cells are commonly identified by using four indices instead of three. The HCP crystal plane indices, called Miller-Bravais indices, are denoted by the letters h, k, i and l and are enclosed in parentheses as ($hkil$). These four-digit hexagonal indices are based on a coordinate system with four axes, as shown in Figure 5.8 in an HCP unit cell. There are three basal axes, a_1, a_2 and a_3, which make 120° with each other. The fourth axis or axis is the vertical axis located at the center of the unit cell. The a unit of measurement along the a_1, a_2 and a_3 axes is the distance between the atoms along these axes and is indicated in Figure 5.8. The unit of measurement along the c-axis is the height of the unit cell. The reciprocals of the intercepts that a crystal plane makes with the a_1, a_2 and a_3 axes give the h, k and i indices, while the reciprocal of the intercept with the c-axis gives the l index.

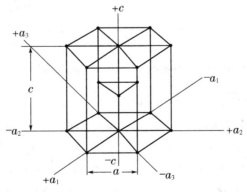

Figure 5.8 The four coordinate axes [a_1, a_2, a_3 and c] of the HCP crystal structure unit cell

(2) Basal planes

The basal planes of the HCP unit cell are very important planes for this unit cell and are indicated in Figure 5.9a. Since the basal plane on the top of the HCP unit cell in Figure 5.9a is parallel to the a_1, a_2 and a_3 axes, the intercepts of this plane with these axes will all be infinite. Thus, $a_1 = \infty$, $a_2 = \infty$ and $a_3 = \infty$. The c-axis, however, is unity since the top basal plane intersects the c axis at unit distance. Taking the reciprocals of these intercepts gives the Miller-Bravais indices for the HCP basal plane. Thus $h = 0$, $k = 0$, $i = 0$ and $l = 1$. The HCP basal plane is, therefore, a zero-zero-zero-one or (0001) plane.

(3) Prism planes

Using the same method, the intercept of the front prism plane (ABCD) of Figure 5.9b are $a_1 = +1$, $a_2 = \infty$, $a_3 = -1$ and $c = \infty$. Taking the reciprocals of these intercepts gives $h = 1$, $k = 0$, $i = -1$ and $l = 0$, or the $(10\bar{1}0)$ plane. Similarly, the ABEF prism plane of Figure 5.9b has the indices $(1\bar{1}00)$ and the DCGH plane the indices $(01\bar{1}0)$. All HCP prism planes can be identified collectively as the $\{10\bar{1}0\}$ family of planes.

Sometimes HCP planes are identified only by three indices (hkl) since $h + k = -i$. However, the ($hkil$) indices are used commonly because they reveal the hexagonal symmetry of the HCP unit cell.

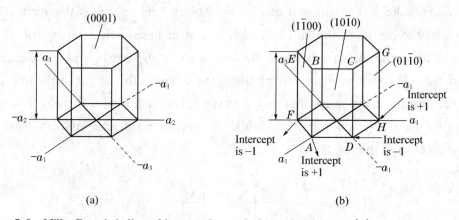

Figure 5.9 Miller-Bravais indices of hexagonal crystal planes: basal planes (a) and prism planes (b)

(4) Direction indices in HCP unit cells

Directions in HCP unit cells are also usually indicated by four indices h, k, i and l, enclosed by square brackets as $[hkil]$. The h, k and i indices are lattice vectors in the a_1, a_2 and a_3 directions, respectively (Figure 5.8), and the l index is a lattice vector in the c direction. To maintain uniformity for both HCP indices for planes and directions, it has been agreed that $h + k = -i$ for directions also, leading to a cumbersome method of designating directions whose method of determination is beyond the scope of this book. Some of the important directions in HCP unit cells are indicated in Figure 5.10.

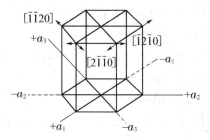

Figure 5.10 Some important directions in HCP unit cell

5.3 Principal Metallic Crystal Structures

Most elemental metals (about 90 percent) crystallize upon solidification into three densely packed crystal structures: body-centered cubic (BCC) (Figure 5.11a) face centered cubic (FCC) (Figure 5.11b) and hexagonal close-packed (HCP) (Figure 5.11c). The HCP structure is a denser modification of the simple hexagonal crystal structure shown in Figure 5.2. Most metals crystallize in these dense-packed[5] structures because energy is released as the atoms come closer together and bond more tightly with each other. Thus, the densely packed structures are in lower and more stable energy arrangements.

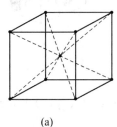 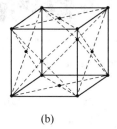 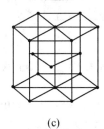

(a) (b) (c)

Figure 5.11 Principal metal crystal structure unit cells: body-centered cubic (a), face-centered cubic (b) and hexagonal close-packed (c)

The extremely small size of the unit cells of crystalline metals which are shown in Figure 5.11 should be emphasized. The cube side of the unit cell of body-centered cubic iron, for example, at room temperature is equal to 0.287×10^{-9} m, or 0.287 nanometer (nm). Therefore, if unit cells of pure iron are lined up side by side, in 1 mm there will be

$$1 \text{ mm} \times \frac{1 \text{unit cell}}{0.287 \text{ nm} \times 10^{-6} \text{mm/nm}} = 3.48 \times 10^6 \text{ unit cells} \qquad (5-2)$$

Let us now examine in detail the arrangement of the atoms in the three principal crystal structure unit cells. Although an approximation, we shall consider atoms in these crystal structures to be hard spheres. The distance between the atoms (interatomic distance) in crystal structures can be deter mined experimentally by X-ray diffraction analysis. For example, the interatomic distance between two aluminum atoms in a piece of pure aluminum at 20 ℃ is 0.286 2 nm. The radius of the aluminum atom in the aluminum metal is assumed to be half the interatomic distance, or 0.143 nm. The atomic radii of selected metals are listed in Tables 5.2 to 5.4.

5.3.1 Body-Centered Cubic (BCC) Crystal Structure

First, consider the atomic-site unit cell for the BCC crystal structure shown in Figure 5.12a. In this unit cell the solid spheres represent the centers where atoms are located and clearly indicate their relative positions. If we represent the atoms in this cell as hard spheres, then the unit cell appears as shown in Figure 5.12b. In this unit cell we see that the central atom is surrounded by eight nearest neighbors and is said to have a coordination number of 8.

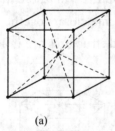

Figure 5.12 BCC unit cells: atomic-site unit cell (a), hard-sphere (b) and isolated unit cell (c)

If we isolate a single hard-sphere[⑥] unit cell, we obtain the model shown in Figure 5.12c. Each of these cells has the equivalent of two atoms per unit cell. One complete atom is located at the center of the unit cell, and an eighth of a sphere is located at each corner of the cell, making the equivalent of another atom. Thus there is a total of 1 (at the center) $+ 8 \times \frac{1}{8}$ (at the corners) = 2 atoms per unit cell. The atoms in the BCC unit contact each other across the cube diagonal, as indicated in Figure 5.13, so that the relationship between the length of the cube side a and the atomic radius R is

$$\sqrt{3}a = 4R \text{ or } a = \frac{4R}{\sqrt{3}} \tag{5-3}$$

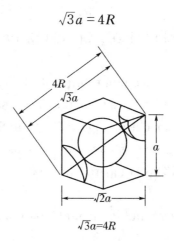

Figure 5.13 BCC unit cell-showing relationship between the lattice constant a and the atomic radius R

If the atoms in the BCC unit cell are considered to be spherical, an atomic packing factor[①](APF) can be calculated by using the equation

$$\text{Atomic packing factor (APF)} = \frac{\text{volume of atoms in unit cell}}{\text{volume of unit cell}} \quad (5-4)$$

Using this equation, the APF for the BCC unit cell (Figure 5.11) is calculated to be 68 percent. That is, 68 percent of the volume of the BCC unit cell is occupied by atoms and the remaining 32 percent is empty space. The BCC crystal structure is not a close-packed structure since the atoms could be packed closer together. Many metals such as iron, chromium, tungsten, molybdenum and vanadium have the BCC crystal structure at room temperature. Table 5.2 lists the lattice constants and atomic radii of selected BCC metals.

Table 5.2 Selected metals which have the BCC crystal structure at room temperature (20 ℃) and their lattice constants and atomic radii

Metal	Lattice constant a/nm	Atomic radius R ∗/nm
Chromium	0.289	0.125
Iron	0.287	0.124
Molybdenum	0.315	0.136
Potassium	0.533	0.231
Sodium	0.429	0.186
Tantalum	0.330	0.143
Tungsten	0.316	0.137
Vanadium	0.304	0.132

∗ Calculated from lattice constants by using Eq. (5-3), $R = \dfrac{\sqrt{3}a}{4}$

5.3.2 Face-Centered Cubic (FCC) Crystal Structure

Consider next the FCC lattice-point unit cell of Figure 5.14a. In this unit cell there is one lattice point at each corner of the cube and one at the center of each cube face. The hard-sphere model of Figure 5.14b indicates that the atoms in the FCC crystal structure are packed as close together as possible. The APF for this close-packed structure is 0.74 as compared to 0.68 for the BCC structure which is not close-packed.

The FCC unit cell as shown in Figure 5.14c has the equivalent of four atoms per unit cell. The eight corner octants account for one atom $\left(8 \times \frac{1}{8} = 1\right)$, and the six half-atoms on the cube faces contribute another three atoms, making a total of four atoms per unit cell. The atoms in the FCC unit cell contact each other across the cubic diagonal, as indicated in Figure 5.15, so that the relationship between the length of the cube side a and the atomic radius R is

$$\sqrt{2}a = 4R \quad \text{or} \quad a = \frac{4R}{\sqrt{2}} \tag{5-5}$$

The APF for FCC crystal structure is 0.74, which is greater than the 0.68 factor for the BCC structure. The APF of 0.74 is for the closest packing possible of "spherical atoms". Many metals such as aluminum, copper, lead, nickel and iron at elevated temperatures (912 to 1394 ℃) crystallize with the FCC crystal structure. Table 5.3 lists the lattice constants and atomic radii for some selected FCC metals.

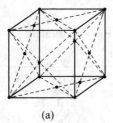 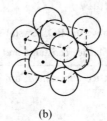 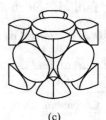

(a) (b) (c)

Figure 5.14 FCC unit cells: atomic-site unit cell (a), hard-sphere unit cell (b) and isolated unit cell (c)

Table 5.3 Selected metals which have the FCC crystal structure at room temperature (20 ℃) and their lattice constants and atomic radii

Metal	Lattice constant a/nm	Atomic radius $R*$/nm
Aluminum	0.405	0.143
Copper	0.361 5	0.128
Gold	0.408	0.144

(to be continued)

Metal	Lattice constant a/nm	Atomic radius $R*$/nm
Lead	0.495	0.175
Nickel	0.352	0.125
Platinum	0.393	0.139
Silver	0.409	0.145

* Calculated from lattice constants by using Eq. (5-5), $R = \dfrac{\sqrt{2}a}{4}$

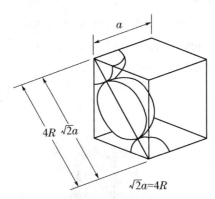

Figure 5.15 FCC unit cell showing relationship between the lattice constant a and atomic radius R. Since the atoms touch across the face diagonals $\sqrt{2}a = 4R$

5.3.3 Hexagonal Close-Packed (HCP) Crystal Structure

The third common metallic crystal structure is the HCP structure shown in Figure 5.16. Metals do not crystallize into the simple hexagonal crystal structure shown in Figure 5.2 because the APF is too low. The atoms can attain a lower energy and a more stable condition by forming the HCP structure of Figure 5.16. The APF of the HCP crystal structure is 0.74, the same as that for the FCC crystal structure since in both structures the atoms are packed as tightly as possible. In both the HCP and FCC crystal structures each atom is surrounded by 12 other atoms, and thus both structures have a coordination number of 12.

The isolated HCP unit cell is shown in Figure 5.16c and has the equivalent of six atoms per unit cell. Three atoms from a triangle in the middle layer, as indicated by the atomic sites in Figure 5.16a. There are six $\dfrac{1}{6}$-atom sections on both the top and bottom layers, making an equivalent of two more atoms $\left(2 \times 6 \times \dfrac{1}{6} = 2\right)$.

Finally, there is one-half an atom in the center of both the top and bottom layers, making the equivalent of one more atom, the total number of atoms in the HCP crystal structure unit cell is thus $3 + 2 + 1 = 6$.

The ratio of the height c of the hexagonal prism of the HCP crystal structure to its basal side a is called the c/a ratio (Figure 5.16a). The c/a ratio for an idea HCP crystal structure consisting of uniform spheres packed as tightly together ratios as possible are 1.633. Table 5.4 lists some important HCP metals and their c/a ratios. Of the metals listed, cadmium and zinc have c/a ratios higher than ideality, which indicates that the atoms in these structures are slightly elongated along the c-axis of the HCP unit cell. The metals magnesium, cobalt, zirconium, titanium and beryllium have c/a ratios less than the ideal ratio. Therefore, in these metals the atoms are slightly compressed in the direction along the c-axis. Thus, for the HCP metals listed in Table 5.4 there is a certain amount of deviation from the ideal hard-sphere model.

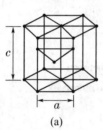
(a)

(b)

(c)

Figure 5.16 HCP unit cells: atomic-site unit cell (a), hard-sphere unit cell (b) and isolated unit cell (c)

Table 5.4 Selected metals which have the HCP crystal structure at room temperature (20 ℃) and their lattice constants, atomic radii and c/a ratios

Metal	Lattice constants / nm		Atomic radius R/nm	c/a ratio	% deviation from ideality
	a	c			
Cadmium	0.297 3	0.561 8	0.149	1.890	+15.7
Zinc	0.266 5	0.494 7	0.133	1.856	+13.6
Ideal HCP	—			1.633	0
Magnesium	0.320 9	0.520 9	0.160	1.623	−0.66
Cobalt	0.250 7	0.469	0.125	1.623	−0.66
Zirconium	0.323 1	0.514 8	0.160	1.593	−2.45
Titanium	0.295 0	0.468 3	0.147	1.587	−2.81
Beryllium	0.228 6	0.358 4	0.113	1.568	−3.98

Chapter 5 Crystal Structure

5.3.4 Comparison of FCC, HCP and BCC Crystal Structures

5.3.4.1 FCC and HCP

As previously pointed out, both the HCP and FCC crystal structures are close-packed structures. That is, their atoms which are considered approximate "spheres" attained, are packed together as closely as possible so that an atomic packing factor of 0.74 is attained. The (111) planes of the FCC crystal structure shown in Figure 5.17a have the identical packing arrangement as the (0001) planes of the HCP crystal structure shown in Figure 5.17b. However, the three-dimensional FCC and HCP crystal structures are not identical because there is a difference in the stacking arrangement of their atomic planes, which can best be described by considering the stacking of hard spheres representing atoms. As a useful analogy, one can imagine the stacking of planes of equal-sized marbles on top of each other, minimizing the space between the marbles.

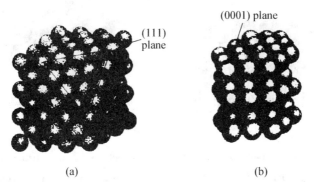

Figure 5.17 Comparison of the FCC crystal structure showing the close-packed [111] planes (a) and the HCP crystal structure, showing the closed-packed [0001] planes (b)

Consider first a plane of close-packed atoms designated the A plane, as shown in Figure 5.18a. Note that there are two different types of empty spaces or voids between the atoms. The voids pointing to the top of the page are designated a voids and those pointing to the bottom of the page, b voids. A second plane of atoms can be placed over the a or b voids and the same three-dimensional structure will be produced. Let us place plane B over the a voids, as shown in Figure 5.18b. Now if a third plane of atoms is placed over plane B to form a closest-packed structure, it is possible to form two different close-packed structures. One possibility is to place the atoms of the third plane in the b voids of the B plane. Then the atoms of this third plane will lie directly over those of the A plane and thus can be designated another A plane (Figure 5.18c). If subsequent planes of atoms are placed in this same alternating stacking arrangement, then the stacking sequence of the three-dimensional structure produced can be denoted by *ABABAB* ... Such a stacking sequence leads to the HCP crystal structure (Figure 5.17b).

The second possibility for forming a simple close-packed structure is to place the third plane in the *a* voids of plane B (Figure 5.18d). This third plane is designated the C plane since its atoms lie neither directly above those of the B plane nor of the A plane. The stacking sequence in this close-packed structure is thus designated *ABABAB* ..., and leads to the FCC structure shown in Figure 5.17a.

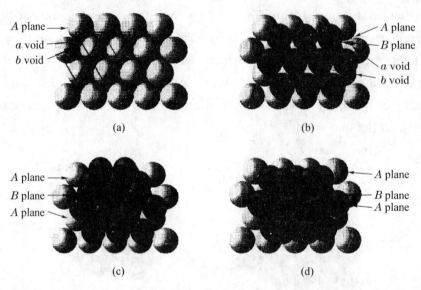

Figure 5.18 Formation of the HCP and FCC crystal structures by the stacking of atomic planes. A plane showing the *a* and *b* voids (a). B plane placed in *a* voids of plane A (b). Third plane placed in *b* voids of B plane, making another A plane and forming the HCP crystal structure (c). Third plane placed in the *a* voids of B plane, making a new C plane and forming the FCC crystal structure (d)

5.3.4.2 BCC

The BCC structure is not a close-packed structure and hence does not have close-packed planes like the {111} planes in the FCC structure and the {0001} planes in the HCP structure. The most densely packed planes in the BCC structure are the {110} family of planes of which the (110) plane is shown in Figure 5.19b. The atoms, however, in the BCC structure do have close-packed directions along the cube diagonals, which are the <111> directions.

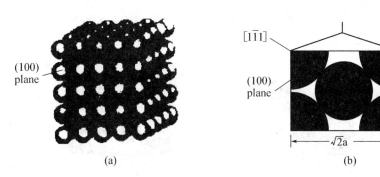

Figure 5.19 BCC crystal structure showing the (100) plane (a) and a section of the (110) plane (b). Note that this is not a close-packed structure, but that diagonals are lose-packed directions

5.3.5 Volume, Planar and Linear Density Unit-Cell Calculations

5.3.5.1 Volume Density

Using the hard-sphere atomic model for the crystal structure unit cell of a metal and a value for the atomic radius of the metal obtained from X-ray diffraction analysis[8], a value for the volume density of a metal can be obtained by using the equation

$$\text{Volume density of metal} = \rho_v = \frac{\text{mass/unit cell}}{\text{volume/unit cell}} \quad (5-6)$$

For example, a value of 8.98 mg/m³ is obtained for the density of copper through the calculation by using the equation (5-6). The handbook experimental value for the density of copper is 8.96 mg/m³. The slightly lower density of the experimental value could be attributed to the absence of atoms at some atomic sites (vacancies), line defects and mismatch where grains meet (grain boundaries). Another cause of the discrepancy could also due to the atoms not being perfect spheres.

5.3.5.2 Planar Atomic Density

Sometimes it is important to determine the atomic densities on various crystal planes. To do this a quantity called the planar atomic density[9] is calculated by using the relationship.

Planar of metal density = ρ_v

$$= \frac{\text{equiv. no. of atoms whose centers are intersected by selected area}}{\text{selected area}} \quad (5-7)$$

For convenience the area of a plane which intersects a unit cell is usually used in these calculations, as shown, for example, in Figure 5.20 for (110) plane in a BCC unit cell. In order for an atom area to be counted in this calculation, the plane of interest must intersect the center of an atom.

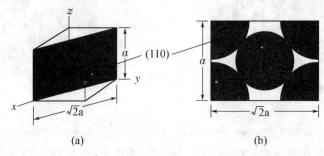

Figure 5.20 A BCC atomic-site unit cell showing a shaded [110] plane (a) and areas of atoms in BCC unit cell cut by the [110] plane (b)

5.3.5.3 Linear Atomic Density

Sometimes it is important to determine the atomic densities in various directions in crystal structures. To do this a quantity called the linear atomic density[⑩] is calculated by using the relationship:

Linear atomic density = ρ_l

$$= \frac{\text{no. of atoms diam. intersected by selected length of line direction of interest}}{\text{selected length of line}}$$

(5-8)

5.3.6 Polymorphism or Allotropy

Many elements and compounds exist in more than one crystalline form under different conditions of temperature and pressure. This phenomenon is termed polymorphism or allotropy. Many industrially important metals such as iron, titanium and cobalt undergo allotropic transformations at elevated temperatures at atom spheric pressure. Table 5.5 lists some selected metals which show allotropic transformations and the structure changes which occur.

Iron exists in both BCC and FCC crystal structures over the temperature range from room temperature to its melting point at 1 539 ℃, as shown in Figure 5.21. α iron exists from −273 to 912 ℃ and has the BCC crystal structure. γ iron exists from 912 to 1 394 ℃ and has FCC crystal structure. δ iron exists from 1 394 to 1 539 ℃, which is the melting point of iron. The crystal structure of α iron is also BCC but with a larger lattice constant than a iron.

Chapter 5 Crystal Structure

Table 5.5 Allotropic crystalline forms of some metals

Metal	Crystal structure at room temperature	At other temperatures / ℃
Ca	FCC	BCC [>447]
Co	HCP	FCC [>447]
Hf	HCP	BCC [>1 742]
Fe	BCC	FCC [912~1 394] BCC [>1 394]
Li	BCC	HCP [<-193]
Na	BCC	HCP [<-233]
Tl	HCP	BCC [>234]
Ti	HCP	BCC [>883]
Y	HCP	BCC [>1481]
Zr	HCP	BCC [>872]

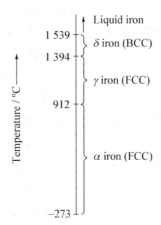

Figure 5.21 Allotropic crystalline forms of iron over temperature ranges at atmospheric pressure

Notes

① unit cell 晶胞,单胞；单位晶格
② space lattice 空间点阵
③ Bravais Lattice 布拉维点阵
④ base-centered 底心的
⑤ dense-packed 密集堆积的
⑥ hard-sphere 硬球
⑦ atomic packing factor 原子堆积因数
⑧ X-ray diffraction analysis X 射线衍射分析
⑨ planar atomic density 平面原子密度
⑩ linear atomic density 线性原子密度

Chapter 5　Crystal Structure

◆ **Vocabulary**

approximation ［ə'prɒksɪ'meɪʃn］ n. ［数］近似法;接近;［数］近似值

basal ［'beɪsəl］ adj. 基部的;基础的

bracket ［'brækɪt］ n. 支架;括号;墙上凸出的托架　vt. 括在一起;把……归入同一类;排除

cadmium ［'kædmɪəm］ n. ［化学］镉

cobalt ［'kəʊbɔːlt］ n. ［化学］钴

coordinate ［kəʊ'ɔːdɪneɪt］ n. 坐标

cumbersome ［'kʌmbəsəm］ adj. 笨重的;累赘的;难处理的

deter ［dɪ'tɜː］ vt. 制止,阻止;使打消念头

deviation ［diːvɪ'eɪʃən］ n. 偏差;误差;背离

diagonal ［daɪ'ægənəl］ n. 对角线;斜线　adj. 斜的;对角线的;斜纹的

diffraction ［dɪ'frækʃn］ n.（光、声等的）衍射,绕射

digit ［'dɪdʒɪt］ n. 数字;手指或足趾;一指宽

discrepancy ［dɪs'krepənsɪ］ n. 不符;矛盾;相差

equispaced ［iːkwɪ'speɪst］ adj. 平均间隔的;均布的

fractional ［'frækʃənəl］ adj. 部分的;［数］分数的,小数的

geometry ［dʒɪ'ɒmɪtrɪ］ n. 几何学;几何结构

hexagonal ［hek'sægənəl］ adj. 六边的,六角形的

interaxial ［ɪntə'æksɪəl］ adj. 轴间的

intercept ［ɪntə'sept］ n. 拦截;［数］截距;截获的情况

interplanar ［ɪntə'pleɪnə］ adj. 晶面间的;平面间的

marble ［'mɑːbl］ n. 大理石;大理石制品;弹珠

mismatch ［'mɪsmætʃ］ n. 错配;不协调　vt. 使配错

monoclinic ［mɒnəʊ'klɪnɪk］ adj. ［晶体］单斜的;［晶体］单斜晶体的

nonparallel ［ˌnɒn'pærəlel］ adj. 非平行的

notation ［nəʊ'teɪʃn］ n. 符号;乐谱;注释;记号法

orthorhombic ［ɔː θəʊ'rɒmbɪk］ adj. ［晶体］正交晶的;斜方晶系的

parentheses ［pə'renθəsɪz］ n. 括号,圆括号,圆括弧

planar ［'pleɪnə］ adj. 平面的;二维的;平坦的

platinum ［'plætɪnəm］ n. ［化学］铂

polymorphism ［ˌpɒlɪ'mɔːfɪzm］ n. 多态性;多形性;同质多晶

potassium ［pə'tæsɪəm］ n. ［化学］钾

prism ［'prɪzm］ n. 棱镜;［晶体］［数］棱柱

reciprocal ［rɪ'sɪprəkəl］ n. ［数］倒数　adj. ［数］倒数的

rectangular ［rek'tæŋgjʊlə］ adj. 矩形的;成直角的

rhombohedral ［rɒmbəʊ'hiːdrəl］ adj. 菱形的,［数］斜方六面体的

tantalum ［'tæntələm］ n. ［化学］钽

tetragonal ［tɪ'trægənl］ adj. ［数］四角形的

triclinic ［traɪ'klɪnɪk］ adj. 三斜晶系的,三斜的

vector ［'vektə］ n. 矢量

Exercises

1. Translate the following Chinese phrases into English

(1) 电子结构　　(2) 晶向　　(3) 三维

(4) 轴长　　(5) 立方晶系　　(6) 晶面

(7) 晶格常数　　(8) 实验值

2. Translate the following English phrases into Chinese

(1) individual atom　　(2) crystal lattice

(3) crystallographic orientation　　(4) hexagonal crystal structure

(5) interatomic distance　　(6) volume density

(7) allotropic transformation

3. Translate the following Chinese sentences into English

(1) 大部分固体(特别是金属和陶瓷)中，原子以有序的方式排列，每个原子与其他原子具有相同数量的临近原子，且其他的相同原子具有相同的原子间距和方向。

(2) 金属具有晶体结构。

(3) 它们是晶态的，这意味着其组成原子按照一种规则、重复的模式堆积在一起。

(4) 一个基本单元余下的空间将被填隙空位或空隙所占据，由此形成八面体和四面体之分。

(5) 众所周知，原子中存在一种粒子叫作电子。

(6) 晶体：组成固体的原子(或离子)在微观上的排列具有长程周期性。

4. Translate the following English sentences into Chinese

(1) Electron microscopy has revolutionized the study of crystal imperfections.

(2) All the atoms in simple lattice are equal in the view of physical, chemical and geometrical area.

(3) If three directions, x, y and z, belonging to different planes, are drawn in a crystal, the spacings between the particles arranged along these directions will in the general case be different (say, a, b and c).

(4) To understand the structure of a material, the type of atoms present, and how the atoms are arranged and bonded must be know.

(5) Small as atoms are, electrons are still smaller.

(6) Bravais lattice: A mathematical abstract, indicating the periodical arrangement in space.

(7) Passing through other lattice points out of one crystal plane, we can get a series of parallel and equidistant crystal planes, and the distributing of lattice points in these crystal planes is uniform. These equidistant crystal planes are called crystal plane fancily.

Chapter 5　Crystal Structure

5. Translate the following Chinese essay into English

肉眼看来,固体表现为由连续分布的刚性体组成。但是,实验证明所有的固体都是由基本单元——原子组成。这些原子都不是随机分布的,而是相互之间以一种高度有序的形式排列而成。这样排列的一群原子被称为晶体。根据原子排列几何位置的不同,存在几种晶体结构。这种晶体结构常常影响固体的物理特性。因此,这方面知识显得非常重要。

6. Translate the following English essay into Chinese

Metal polycrystals are aggregates of numerous tiny cubic crystallites or hexagonal crystallites. Since the anisotropy of crystal lattices leads to the anisotropy of elastic properties for crystals, the elastic properties of a hexagonal polycrystal depend on not only single crystal elastic constants but also the crystalline orientation distribution.

扫一扫,查看更多资料

Chapter 6　Defect Structure

6.1　Introduction

For a crystalline solid we have tacitly assumed that perfect order exists throughout the material on an atomic scale. However, such an idealized solid does not exist; all contain large numbers of various defects or imperfections. As a matter of fact, many of the properties of materials are profoundly sensitive to deviations from crystalline perfection; the influence is not always adverse, and often specific characteristics are deliberately fashioned by the introduction of controlled amounts or numbers of particular defects.

By "crystalline defect" is meant a lattice irregularity having one or more of its dimensions on the order of an atomic diameter. Classification of crystalline imperfections is frequently made according to geometry or dimensionality of the defect. Several different imperfections are discussed in this chapter, including point defects (those associated with one or two atomic positions), linear (or one-dimensional) defects, as well as interfacial defects, or boundaries, which are two-dimensional. Impurities in solids are also discussed, since impurity atoms may exist as point defects.

6.2　Point Defects

6.2.1　Point Defects in Metals

The simplest of the point defects is a vacancy, or vacant lattice site, one normally occupied from which an atom is missing (Figure 6.1). All crystalline solids contain vacancies and, in fact, it is not possible to create such a material that is free of these defects. The necessity of the existence of vacancies is explained using principles of thermodynamics; in essence, the presence of vacancies increases the entropy (i.e., the randomness) of the crystal.

The equilibrium number of vacancies N_v for a given quantity of material depends on and increases with temperature according to

$$N_v = N\exp\left(-\frac{Q_v}{kT}\right) \qquad (6-1)$$

In this expression, N is the total number of atomic sites, Q_v is the energy required for the formation of a vacancy, T is the absolute temperature in kelvins, and k is the gas or Boltzmann's constant[①]. The value of k is 1.38×10^{-23} J/atom·K, or 8.62×10^{-5} eV/atom·K, depending on the units of Q_v. Thus, the number of vacancies increases exponentially with temperature; that is, as Tin Equation 6.1 increases, so does also the expression $\exp(Q_v/kT)$. For most metals, the fraction of vacancies N_v/N just below the melting temperature is on the order of 10^{-4}; that is, one lattice site out of 10000 will be empty. As ensuing discussions indicate, a number of other material parameters have an exponential dependence on temperature similar to that of Equation 6.1.

A self-interstitial[②] is an atom from the crystal that is crowded into an interstitial site, a small void space that under ordinary circumstances is not occupied. This kind of defect is also represented in Figure 6.1. In metals, a self-interstitial introduces relatively large distortions in the surrounding lattice because the atom is substantially larger than the interstitial position in which it is situated. Consequently, the formation of this defect is not highly probable, and it exists in very small concentrations, which are significantly lower than for vacancies.

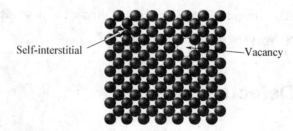

Figure 6.1 Two-dimensional representations of a vacancy and a self-interstitial

6.2.2 Point Defects in Ceramics

Point defects also may exist in ceramic compounds. As with metals, both vacancies and interstitials are possible; however, since ceramic materials contain ions of at least two kinds, defects for each ion type may occur. For example, in

Chapter 6 Defect Structure

NaCl, Na interstitials and vacancies and Cl interstitials and vacancies may exist. It is highly improbable that there would be appreciable concentrations of anion (Cl⁻) interstitials. The anion is relatively large, and to fit into a small interstitial position, substantial strains on the surrounding ions must be introduced. Anion and cation vacancies and a cation interstitial are represented in Figure 6.2.

The expression defect structure is often used to designate the types and concentrations of atomic defects in ceramics. Because the atoms exist as charged ions, when defect structures are considered, conditions of electroneutrality must be maintained. Electroneutrality is the state that exists when there are equal numbers of positive and negative charges from the ions. As a consequence, defects in ceramics do not occur alone. One such type of defect involves a cation-vacancy and a cation-interstitial pair[3]. This is called a Frenkel defect[4] (Figure 6.3). It might be thought of as being formed by a cation leaving its normal position and moving into an interstitial site. There is no change in charge because the cation maintains the same positive charge as an interstitial.

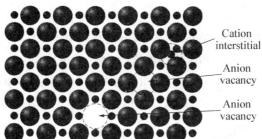

Figure 6.2 Schematic representations of cation and anion vacancies and a cation interstitial

Another type of defect found in AX materials is a cation vacancy-anion vacancy pair known as a Schottky defect[5], also schematically diagrammed in Figure 6.3. This defect might be thought of as being created by removing one cation and one anion from the interior of the crystal and then placing them both at an external surface. Since both cations and anions have the same charge, and since for every anion vacancy there exists a cation vacancy, the charge neutrality of the crystal is maintained.

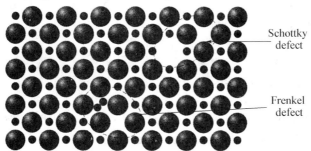

Figure 6.3 Schematic diagram showing Frenkel and Schottky defects in ionic solids

The ratio of cations to anions is not altered by the formation of either a Frenkel or a Schottky defect. If no other defects are present, the material is said to be stoichiometric. Stoichiometry may be defined as a state for ionic compounds wherein there is the exact ratio of cations to anions as predicted by the chemical formula. For example, NaCl is stoichiometric if the ratio of Na^+ ions to Cl^- ions is exactly 1:1. A ceramic compound is nonstoichiometric if there is any deviation from this exact ratio.

Nonstoichiometry may occur for some ceramic materials in which two valences (or ionic) states exist for one of the ion types. Iron oxide (wüstite, FeO) is one such material, for the iron can be present in both Fe^{2+} and Fe^{3+} states; the number of each of these ion types depends on temperature and the ambient oxygen pressure. The formation of an Fe^{3+} ion disrupts the electroneutrality of the crystal by introducing an excess +1 charge, which must be offset by some type of defect. This may be accomplished by the formation of one Fe^{2+} vacancy (or the removal of two positive charges) for every two Fe^{3+} ions that are formed (Figure 6.4). The crystal is no longer stoichiometric because there is one more O ion than Fe ion; however, the crystal remains electrically neutral. This phenomenon is fairly common in iron oxide, and, in fact, its chemical formula is often written as $Fe_{1-x}O$ (where x is some small and variable fraction substantially less than unity) to indicate a condition of nonstoichiometry with a deficiency of Fe.

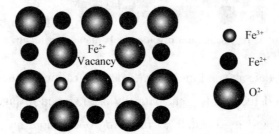

Figure 6.4　Schematic representation of an Fe^{2+} vacancy in FeO that results from the formation of two Fe^{3+} ions

6.2.3　Impurities in Solids

6.2.3.1　Impurities in Metals

A pure metal consisting of only one type of atom just isn't possible; impurity or foreign atoms will always be present, and some will exist as crystalline point defects. In fact, even with relatively sophisticated techniques, it is difficult to refine metals to a purity in excess of 99.9999%. At this level, on the order of 10^{22} to 10^{23} impurity atoms will be present in one cubic meter of material. Most familiar metals are not highly pure; rather, they are alloys, in which impurity atoms have been added intentionally to impart specific characteristics to the material. Ordinarily alloying is used in metals to improve mechanical strength and corrosion resistance. For example, sterling silver is a 92.5% silver − 7.5% copper alloy. In normal ambient environments, pure silver is highly corrosion resistant, but also very soft. Alloying with copper enhances the mechanical strength significantly, without depreciating the corrosion resistance appreciably.

The addition of impurity atoms to a metal will result in the formation of a solid solution and/or a new second phase, depending on the kinds of impurity, their concentrations and the temperature of the alloy.

Several terms relating to impurities and solid solutions deserve mention. With regard to alloys, solute and solvent are terms that are commonly employed. "Solvent" represents the element or compound that is present in the greatest amount; on occasion, solvent atoms are also called host atoms. "Solute" is used to denote an element or compound present in a minor concentration.

6.2.3.2 Solid Solutions

A solid solution forms when, as the solute atoms are added to the host material, the crystal structure is maintained, and no new structures are formed. Perhaps it is useful to draw an analogy with a liquid solution. If two liquids, soluble in each other (such as water and alcohol) are combined, a liquid solution is produced as the molecules intermix, and its composition is homogeneous throughout. A solid solution is also compositionally homogeneous; the impurity atoms are randomly and uniformly dispersed within the solid.

Impurity point defects are found in solid solutions, of which there are two types: substitutional and interstitial. For substitutional, solute or impurity atoms replace or substitute for the host atoms (Figure 6.5). There are several features of the solute and solvent atoms that determine the degree to which the former dissolves in the latter; these are as follows:

(1) Atomic size factor.

Appreciable quantities of a solute may be accommodated in this type of solid solution only when the difference in atomic radii between the two atom types is less than about ±15%. Otherwise the solute atoms will create substantial lattice distortions and a new phase will form.

(2) Crystal structure

For appreciable solid solubility the crystal structures for metals of both atom types must be the same.

(3) Electronegativity

The more electropositive one element and the more electronegative the other, the greater is the likelihood that they will form an intermetallic compound instead of a substitutional solid solution.

(4) Valences

Other factors being equal, a metal will have more of a tendency to dissolve another metal of higher valency than one of a lower valency.

Figure 6.5 Two-dimensional schematic representations of substitutional and interstitial impurity atoms

An example of a substitutional solid solution is found for copper and nickel. These two elements are completely soluble in one another at all proportions. With regard to the aforementioned rules that govern degree of solubility, the atomic radii for copper and nickel are 0.128 and 0.125 nm, respectively, both have the FCC crystal structure, and their electronegativities are 19 and 1.8; finally, the most common valences are +1 for copper (although it sometimes can be +2) and +2 for nickel.

For interstitial solid solutions, impurity atoms fill the voids or interstices among the host atoms (see Figure 6.5). For metallic materials that have relatively high atomic packing factors, these interstitial positions are relatively small. Consequently, the atomic diameter of an interstitial impurity must be substantially smaller than that of the host atoms. Normally, the maximum allowable concentration of interstitial impurity atoms is low (less than 10 %). Even very small impurity atoms are ordinarily larger than the interstitial sites, and as a consequence they introduce some lattice strains on the adjacent host atoms.

Carbon forms an interstitial solid solution when added to iron; the maximum concentration of carbon is about 20 %. The atomic radius of the carbon atom is much less than that for iron: 0.071 nm versus 0.124 nm.

6.2.3.3 Impurities in Ceramics

Impurity atoms can form solid solutions in ceramic materials much as they do in metals. Solid solutions of both substitutional and interstitial types are possible. For an interstitial, the ionic radius of the impurity must be relatively small in comparison to the anion. Since there are both anions and cations, a substitutional impurity will substitute for the host ion to which it is most similar in an electrical sense: if the impurity atom normally forms a cation in a ceramic material, it most probably will substitute for a host cation. For example, in sodium chloride, impurity Ca^{2+} and O^{2-} ions would most likely substitute for Na^+ and Cl^- ions, respectively. Schematic representations for cation and anion substitutional as well as interstitial impurities are shown in Figure 6.6. To achieve any appreciable solid solubility of substituting impurity atoms, the ionic size and charge must be very nearly the same as those of one of the host ions. For an impurity ion having a charge different from the host ion for which it substitutes, the crystal must compensate for this difference in charge so that electroneutrality is maintained with the solid. One way this is accomplished is by the formation of lattice defects-vacancies or interstitials of both ion types, as discussed above.

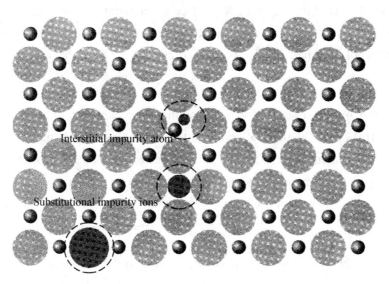

Figure 6.6 Schematic representations of interstitial, anion-substitutional and cation-substitutional impurity atoms in an ionic compound

6.2.4 Point Defects in Polymers

It should be noted that the defect concept is different in polymers (than in metals and ceramics) as a consequence of the chainlike macromolecules and the nature of the crystalline state for polymers. Point defects similar to those found in metals have been observed in crystalline regions of polymeric materials; these include vacancies and interstitial atoms and ions. Chain ends are considered to be defects inasmuch as they are chemically dissimilar to normal chain units; vacancies are also associated with the chain ends. Impurity atoms/ions or groups of atoms/ions may be incorporated in the molecular structure as interstitials; they may also be associated with main chains or as short side branches.

6.2.5 Specification of Composition

It is often necessary to express the composition (or concentration) of an alloy in terms of its constituent elements. The two most common ways to specify composition are weight (or mass) percent and atom percent. The basis for weight percent ($wt\%$) is the weight of a particular element relative to the total alloy weight. For an alloy that contains two hypothetical atoms denoted by 1 and 2, the concentration of 1 in $wt\%$, C_1, is defined as

$$C_1 = \frac{m_1}{m_1 + m_2} \times 100 \qquad (6-2)$$

where m_1 and m_2 represent the weight (or mass) of elements 1 and 2, respectively. The concentration of 2 would be computed in an analogous manner.

The basis for atom percent (at%) calculations is the number of moles of an element in relation to the total moles of the elements in the alloy. The number of moles in some specified mass of a hypothetical element 1, n_{m1}, may be computed as follows:

$$n_{m1} = \frac{m'}{A_1} \qquad (6-3)$$

Here, m'_1 and A_1 denote the mass (in grams) and atomic weight, respectively, for element 1.

Concentration in terms of atom percent of element 1 in an alloy containing 1 and 2 atoms, C'_1, is defined by

$$C'_1 = \frac{n_{m1}}{n_{m1} + n_{m2}} \times 100 \qquad (6-4)$$

In like manner, the atom percent of 2 may be determined.

Atom percent computations also can be carried out on the basis of the number of atoms instead of moles, since one mole of all substances contains the same number of atoms.

6.3 Miscellaneous Imperfections

6.3.1 Dislocations-Linear Defects

A dislocation is a linear or one-dimensional defect around which some of the atoms are misaligned. One type of dislocation is represented in Figure 6.7: an extra portion of a plane of atoms, or half-plane, the edge of which terminates within the crystal. This is termed an edge dislocation; it is a linear defect that centers around the line that is defined along the end of the extra half-plane of atoms. This is sometimes termed the dislocation line, which, for the edge dislocation in Figure 6.7, is perpendicular to the plane of the page. Within the region around the dislocation line there is some localized lattice distortion. The atoms above the dislocation line in Figure 6.7 are squeezed together, and those below are pulled apart; this is reflected in the slight curvature for the vertical planes of atoms as they bend around this extra half-plane. The magnitude of this distortion decreases with distance away from the dislocation line; at positions far removed, the crystal lattice is virtually perfect. Sometimes the edge dislocation in Figure 6.7 is represented by the symbol ⊥, which also indicates the position of the dislocation line. An edge dislocation may also be formed by an extra half-plane of atoms that is included in the bottom portion of the crystal; its designation is a ⊤.

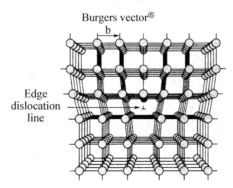

Figure 6.7 The atom positions around an edge dislocation; extra half-plane of atoms shown in perspective

Another type of dislocation, called a screw dislocation[7], exists, which may be thought of as being formed by a shear stress that is applied to produce the distortion shown in Figure 6.8a: the upper front region of the crystal is shifted one atomic distance to the right relative to the bottom portion. The atomic distortion associated with a screw dislocation is also linear and along a dislocation line, line *AB* in Figure 6.8b. The screw dislocation derives its name from the spiral or helical path or ramp that is traced around the dislocation line by the atomic planes of atoms. Sometimes the symbol ↻ is used to designate a screw dislocation.

Most dislocations found in crystalline materials are probably neither pure edge nor pure screw, but exhibit components of both types; these are termed mixed dislocations. All three dislocation types are represented schematically in Figure 6.9; the lattice distortion that is produced away from the two faces is mixed, having varying degrees of screw and edge character.

The magnitude and direction of the lattice distortion associated with a dislocation is expressed in terms of a Burgers vector, denoted by a **b**. Burgers vectors are indicated in Figures 6.7 and 6.8 for edge and screw dislocations, respectively. Furthermore, the nature of a dislocation (*i. e.*, edge, screw or mixed) is defined by the relative orientations of dislocation line and Burgers vector. For an edge, they are perpendicular (Figure 6.7), whereas for a screw, they are parallel (Figure 6.8); they are neither perpendicular nor parallel for a mixed dislocation. Also, even though a dislocation changes direction and nature within a crystal (*e. g.*, from edge to mixed to screw), the Burgers vector will be the same at all points along its line. For example, all positions of the curved dislocation[®] in Figure 6.9 will have the Burgers vector shown. For metallic materials, the Burgers vector for a dislocation will point in a close-packed crystallographic direction and will be of magnitude equal to the interatomic spacing.

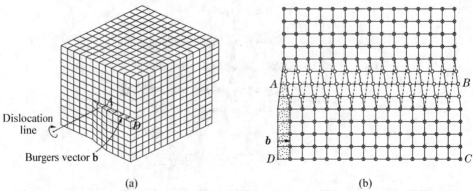

Figure 6.8 A screw dislocation within a crystal(a). The screw dislocation in (a) as viewed from above (b). The dislocation line extends along line *AB*. Atom positions above the slip plane are designated by open circles, those below by solid circles

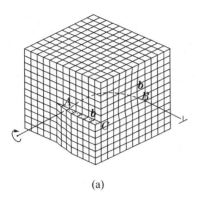
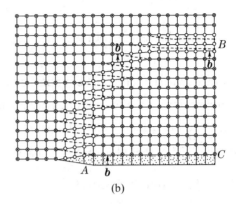

(a)   (b)

Figure 6.9 Schematic representation of a dislocation that has edge, screw, and mixed character(a). Top view, where open circles denote atom positions above the slip plane (b). Solid circles, atom positions below. At point A, the dislocation is pure screw, while at point B, it is pure edge. For regions in between where there is curvature in the dislocation line. The character is mixed edge and screw

Dislocations can be observed in crystalline materials using electron-microscopic techniques. In Figure 6.10, a high-magnification transmission electron micrograph, the dark lines are the dislocations.

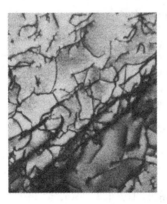

Figure 6.10 A transmission electron micrograph of a titanium alloy in which the dark lines are dislocations. 51,450×

Virtually all crystalline materials contain some dislocations that were introduced during solidification, during plastic deformation, and as a consequence of thermal stresses that result from rapid cooling. Dislocations are involved in the plastic deformation of these materials. Dislocations have been observed in polymeric materials; however, some controversy exists as to the nature of dislocation structures in polymers and the mechanism(s) by which polymers plastically deform.

6.3.2 Interfacial Defects

Interfacial defects are boundaries that have two dimensions and normally separate regions of the materials that have different crystal structures and/or crystallographic orientations. These imperfections include external surfaces, grain boundaries, twin boundaries, stacking faults[9] and phase boundaries.

6.3.2.1 External Surfaces

One of the most obvious boundaries is the external surface, along which the crystal structure terminates. Surface atoms are not bonded to the maximum number of nearest neighbors, and are therefore in a higher energy state than the atoms at interior positions. The bonds of these surface atoms that are not satisfied give rise to a surface energy, expressed in units of energy per unit area (J/m^2 or erg/cm^2). To reduce this energy, materials tend to minimize, if at all possible, the total surface area. For example, liquids assume a shape having a minimum area-the droplets become spherical. Of course, this is not possible with solids, which are mechanically rigid.

6.3.2.2 Grain Boundaries

Another interfacial defect, the grain boundary, was introduced as the boundary separating two small grains or crystals having different crystallographic orientations in polycrystalline materials. A grain boundary is represented schematically from an atomic perspective in Figure 6.11. Within the boundary region, which is probably just several atom distances wide, there is some atomic mismatch in a transition from the crystalline orientation of one grain to that of an adjacent one.

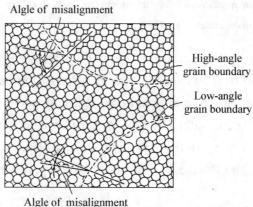

Figure 6.11 Schematic diagram showing low-and high-angle grain boundaries and the adjacent atom positions

Various degrees of crystallographic misalignment between adjacent grains are possible (Figure 6.11). When this orientation mismatch is slight, on the order of a few degrees, then the term small- (or low-) angle grain boundary is used. These boundaries can be described in terms of dislocation arrays. One simple small-angle grain boundary is formed when edge dislocations are aligned in the manner of Figure 6.12. This type is called a tilt boundary; the angle of misorientation, θ, is also indicated in the figure. When the angle of misorientation is parallel to the boundary, a twist boundary result, which can be described by an array of screw dislocations.

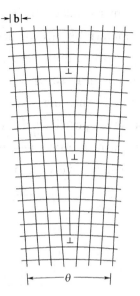

Figure 6.12 Demonstration of how a tilt boundary having an angle of misorientation θ results from an alignment of edge dislocations

The atoms are bonded less regularly along a grain boundary (e. g., bond angles are longer), and consequently, there is an interfacial or grain boundary energy similar to the surface energy described above. The magnitude of this energy is a function of the degree of misorientation, being larger for high-angle boundaries. Grain boundaries are more chemically reactive than the grains themselves as a consequence of this boundary energy. Furthermore, impurity atoms often preferentially segregate along these boundaries because of their higher energy state. The total interfacial energy is lower in large or coarse-grained materials than in fine-grained ones, since there is less total boundary area in the former. Grains grow at elevated temperatures to reduce the total boundary energy.

In spite of this disordered arrangement of atoms and lack of regular bonding along grain boundaries, a polycrystalline material is still very strong; cohesive forces within and across the boundary are present. Furthermore, the density of a polycrystalline specimen is virtually identical to that of a single crystal of the same material.

6.3.2.3 Twin Boundaries

A twin boundary is a special type of grain boundary across which there is a specific mirror lattice symmetry; that is, atoms on one side of the boundary are located in mirror image positions of the atoms on the other side (Figure 6.13). The region of material between these boundaries is appropriately termed a twin. Twins result from atomic displacements that are produced from applied mechanical shear forces (mechanical twins), and also during annealing heat treatments following deformation (annealing twins). Twinning occurs on a definite crystallographic plane and in a specific direction, both of which depend on the crystal structure. Annealing twins are typically found in metals that have the FCC crystal structure, while mechanical twins are observed in BCC and HCP metals.

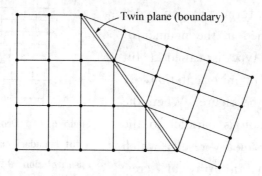

Figure 6.13　Schematic diagram showing a twin plane or boundary and the adjacent atom positions (dark circles)

6.3.2.4　Miscellaneous Interfacial Defects

Other possible interfacial defects include stacking faults, phase boundaries and ferromagnetic domain walls[9]. Stacking faults are found in FCC metals when there is an interruption in the ABCABCABC ... stacking sequence of close-packed planes. Phase boundaries exist in multiphase materials across which there is a sudden change in physical and/or chemical characteristics.

With regard to polymeric materials, the surfaces of chain-folded layers are considered to be interfacial defects, as are boundaries between two adjacent crystalline regions.

Associated with each of the defects discussed in this section is an interfacial energy, the magnitude of which depends on boundary type, and which will vary from material to material. Normally, the interfacial energy will be greatest for external surfaces and least for domain walls.

6.3.3　Bulk or Volume Defects

Other defects exist in all solid materials that are much larger than those heretofore discussed. These include pores, cracks, foreign inclusions and other phases. They are normally introduced during processing and fabrication steps. Some of these defects and their effects on the properties of materials are discussed in relevant chapters.

6.3.4 Atomic Vibrations

Every atom in a solid material is vibrating very rapidly about its lattice position within the crystal. In a sense, these vibrations may be thought of as imperfections or defects. At any instant of time not all atoms vibrate at the same frequency and amplitude, nor with the same energy. At a given temperature there will exist a distribution of energies for the constituent atoms about an average energy. Over time the vibrational energy of any specific atom will also vary in a random manner. With rising temperature, this average energy increases, and, in fact, the temperature of a solid is really just a measure of the average vibrational activity of atoms and molecules. At room temperature, a typical vibrational frequency is on the order of 10^{13} vibrations per second, whereas the amplitude is a few thousandths of a nanometer.

Many properties and processes in solids are manifestations of this vibrational atomic motion. For example, melting occurs when the vibrations are vigorous enough to rupture large numbers of atomic bonds.

Notes

① Boltzmann's constant 玻尔兹曼常数
② self-interstitial 自间隙
③ cation-interstitial pair 间隙阳离子对
④ Frenkel defect 弗兰克尔缺陷
⑤ Schottky defect 肖特基缺陷
⑥ Burgers vector 伯格斯矢量
⑦ screw dislocation 螺旋位错
⑧ curved dislocation 弯曲的位错
⑨ ferromagnetic domain walls 铁磁畴壁

Vocabulary

ambient [ˈæmbɪənt] n. 周围环境 adj. 周围的;外界的;环绕的
analogous [əˈnæləɡəs] adj. 类似的,相似的
anion [ˈænɪən] n. 阴离子
chainlike [tʃeɪnˈlaɪk] n. 链;束缚;枷锁 (chain 的变形)

chloride [ˈklɔːraɪd] n. 氯化物
deviation [ˌdiːvɪˈeɪʃən] n. 偏差;误差;背离
dimensionality [dɪˌmenʃəˈnælətɪ] n. 维度
dislocation [ˌdɪsləˈkeɪʃən] n. 转位;混乱
droplet [ˈdrɒplɪt] n. 小滴,微滴

Chapter 6 Defect Structure

electroneutrality ［ɪˌlektrəʊˌnuːˈtrælɪtɪ］ n．电中性
entropy ［ˈentrəpɪ］ n．［热］熵（热力学函数）
helical ［ˈhelɪkəl］ adj．螺旋形的
hypothetical ［ˌhaɪpəˈθetɪkəl］ adj．假设的；爱猜想的
intermix ［ˌɪntəˈmɪks］ vt．使……混杂；使……混合 vi．混杂；混合
Kelvins ［ˈkelvɪn］ n．［物理学］开氏度，开尔文（开氏温标的温度单位）
macromolecules ［ˌmækrɒməˈlekjuːlz］ n．［化学］大分子；［高分子］高分子；巨分子（macromolecule 的复数）
miscellaneous ［ˌmɪsəˈleɪnɪəs］ adj．混杂的，各种各样的；多方面的
misorientation ［ˈmɪsˌɔːrɪenˈteɪʃən］ n．取向错误；极向错误；错向
ramp ［ræmp］ n．斜坡，坡道
stoichiometric ［ˌstɔɪkɪəˈmetrɪk］ adj．化学计量的；化学计算的
stoichiometry ［ˌstɔɪkɪˈɒmɪtrɪ］ n．化学计量学
tilt ［tɪlt］ n．倾斜   vt．使倾斜
vibration ［vaɪˈbreɪʃən］ n．振动

Exercises

1. Translate the following Chinese phrases into English

(1) 绝对温度   (2) 点缺陷   (3) 化学式
(4) 无序排列   (5) 机械剪切力   (6) 热处理
(7) 化学特征   (8) 振动能量   (9) 晶体缺陷

2. Translate the following English phrases into Chinese

(1) plastic deformation   (2) thermal stress
(3) twin boundaries   (4) external surface
(5) atomic displacement   (6) multiphase material
(7) impurity atoms   (8) planar defect

3. Translate the following Chinese sentences into English

(1) 加工条件如何影响材料的缺陷结构，以及缺陷如何影响材料的性能，掌握这些内容可用于确定恰当的加工方式以便获得需要的性能，且可用来判断加工条件对性能的影响。

(2) 每个结晶部位都产生一个晶粒，在凝固结束时与邻近的晶粒相会于晶粒边界。

(3) 铸铁缺陷是多种多样的。

(4) 滑移和形成孪晶是最重要的变形方式，并且是造成断裂前塑性变形的主要原因。

(5) 空穴和正电子类似，是由缺少价电子引起的。

4. Translate the following English sentences into Chinese

(1) Controlling the number of dislocations in material by forming the object and by heat treatment is an important tool for the control of the properties of the

material.

(2) Introducing these defects and controlling their number and arrangement provide a powerful tool for the materials engineer to design and create new materials with desirable combinations of properties.

(3) The electronic arrangement influences how the atoms are bonded to one another and helps determine the type of material-metal, ceramic or polymer.

(4) An ideal crystal is constructed by the infinite repetition of identical structural units, and can be separated into the lattice and basis.

5. Translate the following Chinese essay into English

晶体中的格点表示原子的平衡位置,晶格振动便是指原子在格点附近的振动。研究晶格振动式是研究固体宏观和微观过程的重要基础。对晶体热学性质、电学性质、光学性质、超导电性、磁性、结构相变等一系列物理问题,晶格振动都有很重要意义。

6. Translate the following English essay into Chinese

Point defect is the most simple crystal defects, it is a defect that deviation from the crystal structure normal alignment in the node or the adjacent area. Point defect occurs in the crystal lattice constant of one or several range, which is characterized in' the three-dimensional direction are very small in size, for example the vacancy, interstitial atoms, impurity atoms, which can also be called zero dimensional defects. Point defect and temperature are closely related and also known as thermal defects.

扫一扫,查看更多资料

Chapter 7　Structure of Bulk Phase

7.1　Introduction

Structure of bulk phase includes mainly single crystals, polycrystal and amorphism. With the development of the structure of bulk phase, another structure of bulk phase (namely, quasicrystal) was discovered by Shechtman, Blech, Gratias and Cahn in 1984. Therefore, the definition, the concept of higher dimensional space, types of quasicrystals and diffraction pattern symmetries of quasicrystals will be introduced in detail in the chapter.

7.2　Single Crystals

For a crystalline solid, when the periodic and repeated arrangement of atoms is perfect or extends throughout the entirety of the specimen without interruption, the result is a single crystal. All unit cells interlock in the same way and have the same orientation. Single crystals exist in nature, but they may also be produced artificially. They are ordinarily difficult to grow, because the environment must be carefully controlled.

If the extremities of a single crystal are permitted to grow without any external constraint, the crystal will assume a regular geometric shape having flat faces, as with some of the gem stones; the shape is indicative of the crystal structure. A photograph of several single crystals is shown in Figure 7.1. Within the past few years, single crystals have become extremely important in many of our modern technologies, in particular electronic microcircuits, which employ single crystals of silicon and other semiconductors.

Chapter 7　Structure of Bulk Phase

Figure 7.1　Photograph of a rock containing three crystals of pyrite (FeS$_2$). The crystal structure of pyrite is simple cubic and this is reflected in the cubic symmetry of its natural crystal facets

7.3　Polycrystalline Materials

Most crystalline solids are composed of a collection of many small crystals or grains; such materials are termed polycrystalline. Various stages in the solidification of a polycrystalline specimen are represented schematically in Figure 7.2. Initially,

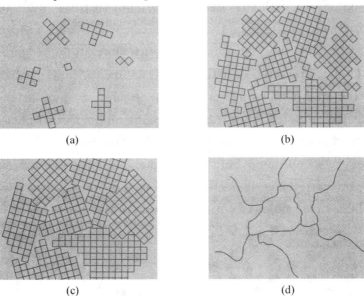

Figure 7.2　Schematic diagrams of the various stages in the solidification of a polycrystalline material; the square grids depict unit cells. Small crystallite nuclei (a). Growth of the crystallites; the obstruction of some grains that are adjacent to one another is also shown(b). Upon completion of solidification, grains having irregular shapes have formed(c). The grain structure as it would appear under the microscope; dark lines are the grain boundaries(d)

small crystals or nuclei form at various positions. These have random crystallographic orientations, as indicated by the square grids. The small grains grow by the successive addition from the surrounding liquid of atoms to the structure of each. The extremities of adjacent grains impinge on one another as the solidification process approaches completion. As indicated in Figure 7.2, the crystallographic orientation varies from grain to grain. Also, there exists some atomic mismatch within the region where two grains meet; this area, called a grain boundary.

7.4 Noncrystalline Solids

It has been mentioned that noncrystalline solids lack a systematic and regular arrangement of atoms over relatively large atomic distances. Sometimes such materials are also called amorphous (meaning literally without form), or supercooled liquids, inasmuch as their atomic structure resembles that of a liquid. An amorphous condition may be illustrated by comparison of the crystalline and noncrystalline structures of the ceramic compound silicon dioxide (SiO_2), which may exist in both states. Figure 7.3a and Figure 7.3b present two-dimensional schematic diagrams for both structures of SiO_2, in which the SiO_4^{4-} tetrahedron is the basic unit (Figure 7.4). Even though each silicon ion bonds to four oxygen ions for both states, beyond this, the structure is much more disordered and irregular for the noncrystalline structure. Whether a crystalline or amorphous solid forms depends on the ease with which a random atomic structure in the liquid can transform to an ordered state during solidification. Amorphous materials, therefore, are characterized by atomic or molecular structures that are relatively complex and become ordered only with some difficulty. Furthermore, rapidly cooling through the freezing temperature favors the formation of a noncrystalline solid, since little time is allowed for the ordering process.

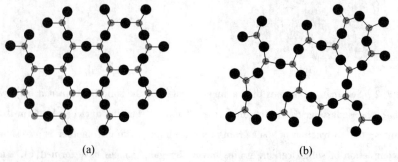

Figure 7.3 Two-dimensional schemes of the structure of crystalline silicon dioxide (a) and noncrystalline silicon dioxide (b). The gray balls and black balls represent Si and O atoms, respectively

Chapter 7 Structure of Bulk Phase

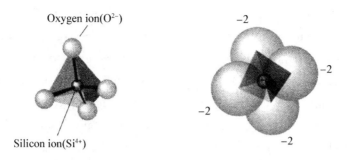

Figure 7.4 A silicon-oxygen (SiO_4^{4-}) tetrahedron

Metals normally form crystalline solids; but some ceramic materials are crystalline, whereas others (i.e., the silica glasses) are amorphous. Polymers may be completely noncrystalline and semicrystalline consisting of varying degrees of crystallinity.

Silicon dioxide (or silica, SiO_2) in the noncrystalline state is called fused silica, or vitreous silica; again, a schematic representation of its structure is shown in Figure 7.3b. Other oxides (e.g., B_2O_3 and GeO_2) may also form glassy structures; these materials, as well as SiO_2, are network formers.

The common inorganic glasses that are used for containers, windows, and so on, are silica glasses to which have been added other oxides such as CaO and Na_2O. These oxides do not form polyhedral networks. Rather, their cations are incorporated within and modify the SiO_4^{-4} network; for this reason, these oxide additives are termed network modifiers. For example, Figure 7.5 is a schematic

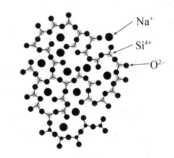

Figure 7.5 Schematic representations of ion positions in a sodium-silicate glass

representation of the structure of a sodium-silicate glass. Still other oxides, such as TiO_2 and Al_2O_3, while not network formers, substitute for silicon and become part of and stabilize the network; these are called intermediates. From a practical perspective, the addition of these modifiers and intermediates lowers the melting point and viscosity of a glass, and makes it easier to form at lower temperatures.

7.5 Quasicrystals

(1) Definition of quasicrystals

In classical crystallography a crystal is defined as a three dimensional periodic

Chapter 7 Structure of Bulk Phase

arrangement of atoms with translational periodicity along its three principal axes. Thus it is possible to obtain an infinitely extended crystal structure by aligning building blocks called unit-cells until the space is filled up. Normal crystal structures can be described by one of the 230 space groups, which describe the rotational and translational symmetry elements present in the structure. Diffraction patterns of these normal crystals therefore show crystallographic point symmetries (belonging to one of the 11 Laue-groups[①]). In 1984, however, Shechtman, Blech, Gratias & Cahn published a paper which marked the discovery of quasicrystals. They showed electron diffraction patterns of an Al-Mn alloy with sharp reflections and 10-fold symmetry. The whole set of diffraction patterns revealed an icosahedral symmetry of the reciprocal space. Since then many stable and meta-stable[②] quasicrystals were found. These are often binary or ternary intermetallic alloys with aluminium as one of the constituents. The icosahedral quasicrystals form one group and the polygonal quasicrystals another (8, 10, 12-fold symmetry). We can state that quasicrystals are materials with perfect long-range order[③], but with no three-dimensional translational periodicity. The former is manifested in the occurrence of sharp diffraction spots and the latter in the presence of a non-crystallographic rotational symmetry.

(2) The concept of higher dimensional space

Since quasicrystals lost periodicity in at least one dimension it is not possible to describe them in 3D-space as easily as normal crystal structures. Thus it becomes more difficult to find mathematical formalisms for the interpretation and analysis of diffraction data. For normal crystals we can assign three integer values (Miller indices) to label the observable reflections. This is due to the three-dimensional translational periodicity of the structure. In order to assign integer indices to the diffraction intensities of quasicrystals, however, at least 5 linearly independent vectors are necessary. So we need 5 indices for polygonal quasicrystals and 6 indices for icosahedral quasicrystals. We can call them generalized Miller indices. The necessary n vectors span an nD-reciprocal space. Therefore there is also an nD-direct space in which a structure can be built that gives rise to a diffraction pattern as it is observed for quasicrystals. To put it simply we can say that in the higher-dimensional space we can describe a quasiperiodic structure as a periodic one. The actual quasiperiodic structure in the 3D-physical space can then be obtained by appropriate projection/section techniques. Thus it is enough to define a single unit cell of the nD-structure. The contents of that nD-unit cell consists of "hyperatoms" (occupation domains, ...) in analogy to the atoms in a normal unit cell. This enables us to describe the whole quasicrystal structure with a finite set of parameters. If we

Chapter 7 Structure of Bulk Phase

described it in 3D-space only, we needed thousands of atoms to obtain a representative volume segment of the whole structure as well as all parameters that go with it (e. g., thousands of positions).

In order to elucidate this concept with a simple example lets have a look at a 1D-"quasicrystal" in form of a Fibonacci chain[④], which is a quasiperiodic sequence of short (dark) and long (gray) segments (Figure 7.6). We embed this 1D-"quasicrystal" in a 2D-"higher dimensional space" which in this case has the form of a simple square lattice. One unit cell of the higher-dimensional space is filled yellow. The axes show the orientation of the two orthonormal subspaces Ve, VI. The slope of Ve with respect to the 2D-lattice has to be an irrational one. In this case it is tau = 1.618 ... Figure 7.6a demonstrates the projection method, where we have a strip of projection with finite width. All points of the 2D-lattice inside this strip are projected onto the external space Ve, thus giving the quasiperiodic sequence (dark, gray.). Figure 7.6b shows the section method in which a hyperplane (here a 1D-line) that is parallel to Ve cuts the higher-dimensional space. The occupation domains that are attached to each of the lattice points (here: bars) intersect with the hyperplane (here: line) thus producing the same quasiperiodic sequence as in Figure 7.6a. These occupation domains (here: bars) extend parallel to the internal space VI. We can generalize this to nD cases with n = 5, 6, ... In that case the internal space would be (n-3)-dimensional and the occupation domains would be two-dimensional (e. g., polygons) or three-dimensional (e. g., polyhedra).

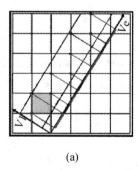
(a)

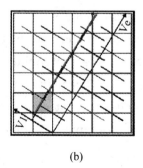
(b)

Figure 7.6　1D-quasiperiodic sequence (Fibonacci chain)

(3) What do We need tilings for

Before quasicrystals were discovered in 1984 the british mathmatician Roger Penrose devised a way to cover a plane in a nonperiodic fashion using two different types of tiles. An example can be seen on the left. The tiles (rhombii) are arranged in a way that they obey certain matching rules. An equivalent tiling can be obtained for a 3D-arrangement. This is called a 3D-Penrose Tiling, which is made up of rhombohedrons instead of the rhombii. Such 2D and 3D-tilings have several important properties, such as the selfsimilarity, which means that any part of the tiling occurs again within a predictable area (or volume). After the discovery of quasicrystals in 1984 a close resemblance was noted between the icosahedral quasicrystal and the 3D-Penrose pattern. By putting atoms at the vertices of a 3D-Penrose pattern one can obtain a Fourier Transform which explains very well the diffraction patterns of the found Al-Mn quasicrystal. In a similar way one can use 2D-Penrose Tilings (Figure 7.7) to approximate a decagonal quasicrystal, which in a simple case consists of two layers with local 5-fold symmetry, which are rotated by 18 degrees so that the projection along the rotation axis gives a 10-fold symmetry. As stated above it is also possible to

Figure 7.7　2D-penrose tilings

derive the vertices of such tilings using the nD-space approach (n>3). In this case we can obtain such a tiling by a projection of an nD periodic lattice (e. g., hypercubic lattice).

(4) Types of quasicrystals

Up to now, quasicrystals may be grouped into four types as follows: 1) Octagonal quasicrystals with local 8-fold symmetry. 2) Decagonal quasicrystals with local 10-fold symmetry. 3) Dodecagonal quasicrystals with local 12-fold symmetry. 4) Icosahedral quasicrystals (axes: 12 × 5-fold, 20 × 3-fold, 30 × 2-fold). 5) "Icosahedral" quasicrystal with broken symmetry.

(5) Diffraction pattern symmetries

The symmetry that determines the type of the quasicrystal is first seen in its diffraction pattern. Some simulations of diffraction patterns are shown in Figure 7.8, which could represent either electron diffraction patterns or the zeroth layers of precession photographs (X-ray):

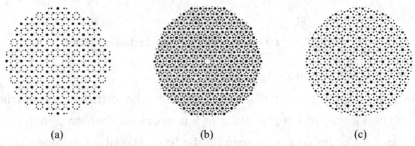

Figure 7.8　Some simulations of diffraction patterns: (a)—octagonal QC, (b)—decagonal QC and (c)—dodecagonal QC

There is a simulation for a Laue pattern (X-ray) from an icosahedral quasicrystal, whereby the X-ray beam is along one of the five-fold axes (Figure 7.9).

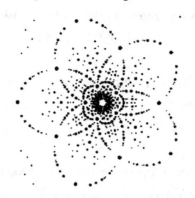

Figure 7.9   A simulation for a Laue pattern (X-ray) from an icosahedral quasicrystal (icosahedral QC)

◆ Notes

① Laue-groups 劳厄群
② meta-stable 亚稳态
③ long-range order 长程有序,长程序
④ Fibonacci chain 斐波纳契链

◆ Vocabulary

decagonal [dɪˈkægənəl] *adj.* 十边形的
dioxide [daɪˈɒksaɪd] *n.* 二氧化物
dodecagonal [ˌdəʊdeˈkægənəl] *adj.* 十二边形的;十二角形的
elucidate [ɪˈluːsɪdeɪt] *vt.* 阐明;说明
hypercubic [ˈhaɪpəkjuːbɪk] *n.* [数]超立方体
hyperplane [ˈhaɪpəpleɪn] *n.* [数]超平面
impinge [ɪmˈpɪndʒ] *vi.* 撞击;侵犯 *vt.* 撞击
interlock [ˌɪntəˈlɒk] *v.* [计]互锁;连锁
microcircuit [ˈmaɪkrəʊˌsɜːkɪt] *n.* [电子]微电路
periodicity [ˌpɪərɪəˈdɪsɪtɪ] *n.* [数]周期性;频率;定期性
polyhedra [ˌpɒlɪˈhiːdrə] *n.* 多面体

quasicrystal [ˈkweɪzaɪˈkrɪstl] *n.* [晶体]准晶体
quasiperiodic [kˈwɑːziːpɪərɪˈɒdɪk] *adj.* 准周期的
rotational [rəʊˈteɪʃənəl] *adj.* 转动的;回转的;轮流的
selfsimilarity [ˈselfsɪməlˈærɪtɪ] *n.* 自相似性
semicrystalline [ˌsemɪˈkrɪstəlaɪn] *n.* 半晶质
silica [ˈsɪlɪkə] *n.* 二氧化硅;[材]硅土
stabilize [ˈsteɪbəlaɪz] *vt.* 使稳固,使安定 *vi.* 稳定,安定
tau [tɔː] *n.* 希腊语的第十九个字母
translational [trænzˈleʃənl] *adj.* 平移的,直移的
zeroth [ˈzɪərəʊθ] *adj.* [数]零的

Chapter 7 Structure of Bulk Phase

◆ Exercises

1. Translate the following Chinese phrases into English

(1) 外部约束 　　　(2) 常规的几何形状 　　　(3) 过冷液体

(4) 多晶材料 　　　(5) 电子衍射花样 　　　(6) 体相结构

2. Translate the following English phrases into Chinese

(1) noncrystalline solids 　　　　　(2) square grid

(3) freezing temperature 　　　　　(4) single crystal

(5) flat face

3. Translate the following Chinese sentences into English

(1) 理论和实验上研究了准晶及其他光子晶体的负折射和近场成像特性。

(2) 这种结构叫作金属键自由电子模型。

(3) 对单晶电子衍射图像的分析方法作了介绍。

(4) 当它们吸收能量的时候，它们的电子跃迁到高能级层，积聚在某一个亚稳态层。

(5) XRD分析表明室温时此系列多晶样品都是单相。

4. Translate the following English sentences into Chinese

(1) These correspond to composites and can be of different chemical compounds or of different phases of the same compound.

(2) Under very high pressure, the samples switched from their amorphous, glassy state to form a single crystal.

(3) The samples showed typical properties of supercooled liquid in the supercooled liquid region.

(4) Two new view points were used when we analyze the structure of the quasicrystal approximant: the icosahedra atom cluster and the layer structure.

5. Translate the following Chinese essay into English

至今，已经发现的准晶有四面体、立方体、十二面体以及五边、八边、十边和十二边棱形对称性。它们也具有独特的、目前正被鼎力研究的物理性质。特别值得提及的是，尽管它们是二元或三元合金，但是它们是非常差的导电体和导热体。

6. Translate the following English essay into Chinese

An understanding of how processing conditions affect defect the structure of a material andhow (and which) defects affect properties can be used to determine the appropriate processing path to obtained the desired properties or how operating conditions will affect properties.

扫一扫，查看更多资料

Part III
Foundation of Material Properties

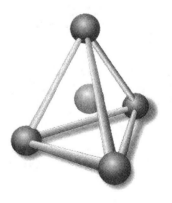

Part III

Foundation of Material Properties

Chapter 8　Mechanical Properties of Materials

8.1　Introduction

Many materials, when in service, are subjected to forces or loads; examples include the aluminum alloy from which an airplane wing is constructed and the steel in an automobile axle. In such situations it is necessary to know the characteristics of the material and to design the member from which it is made such that any resulting deformation will not be excessive and fracture will not occur. The mechanical behavior of a material reflects the relationship between its responseor deformation to an applied load or force. Important mechanical properties are strength, hardness, ductility and stiffness.

The mechanical properties of materials are ascertained by performing carefully designed laboratory experiments that replicate as nearly as possible the service conditions. Factors to be considered include the nature of the applied load and its duration, as well as the environmental conditions. It is possible for the load to be tensile, compressive or shear, and its magnitude may be constant with time, or it may fluctuate continuously. Application time may be only a fraction of a second, or it may extend over a period of many years. Service temperature may be an important factor.

Mechanical properties are of concern to a variety of parties (e. g., producers and consumers of materials, research organizations, and government agencies) that have differing interests. Consequently, it is imperative that there be some consistency in the manner in which tests are conducted, and in the interpretation of their results. This consistency is accomplished by using standardized testing techniques. Establishment and publication of these standards are often coordinated by professional societies. In the United States the most active organization is the American Society for Testing and Materials (ASTM). Its Annual Book of ASTM

Chapter 8 Mechanical Properties of Materials

Standards comprises numerous volumes, which are issued and updated yearly; a large number of these standards relate to mechanical testing techniques.

The role of structural engineers is to determine stresses and stress distributions within members that are subjected to well-defined① loads. This may be accomplished by experimental testing techniques and/or by theoretical and mathematical stress analyses. These topics are treated in traditional stress analysis and strength of materials texts.

Materials and metallurgical engineers, on the other hand, are concerned with producing and fabricating materials to meet service requirements as predicted by these stress analyses. This necessarily involves an understanding of the relationships between the microstructure (i. e., internal features) of materials and their mechanical properties.

Materials are frequently chosen for structural applications because they have desirable combinations of mechanical characteristics. This chapter discusses the stress-strain② behaviors of metals, ceramics, polymers and the related mechanical properties; it also examines their other important mechanical characteristics.

8.2 Concepts of Stress and Strain

If a load is static or changes relatively slowly with time and is applied uniformly over a cross section or surface of a member, the mechanical behavior may be ascertained by a simple stress-strain test; these are most commonly conducted for metals at room temperature. There are three principal ways in which a load may be applied: namely, tension, compression and shear (Figures 8.1a, 8.1b, and 8.1c). In engineering practice many loads are torsional rather than pure shear; this type of loading is illustrated in Figure 8.1d.

8.2.1 Tension Tests

One of the most common mechanical stress-strain tests is performed in tension. As will be seen, the tension test can be used to ascertain several mechanical properties of materials that are important in design. A specimen is deformed, usually to fracture, with a gradually increasing tensile load that is applied uniaxially along the long axis of a specimen. A standard tensile specimen is shown in Figure 8.2. Normally, the cross section is circular, but rectangular specimens are also used. During testing, deformation is confined to the narrow center region,

Chapter 8 Mechanical Properties of Materials

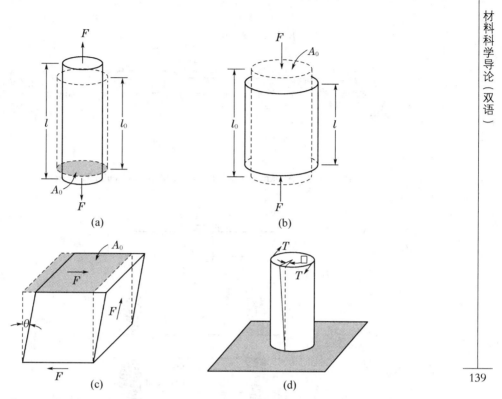

Figure 8.1 Schematic illustration of how a tensile load produces an elongation and positive linear strain. Dashed lines represent the shape before deformation; solid lines, after deformation (a). Schematic illustration of how a compressive load produces contraction and a negative linear strain (b). Schematic representation of shear strain γ, where $\gamma = \tan \theta$ (c). Schematic representation of torsional deformation (i. e., angle of twist) produced by an applied torque T (d)

which has a uniform cross section along its length. The standard diameter is approximately 12.8 mm, whereas the reduced section length should be at least four times this diameter; 60 mm is common. Gauge length is used in ductility computations; the standard value is 50 mm. The specimen is mounted by its ends into the holding grips of the testing apparatus (Figure 8.3). The tensile testing machine is designed to elongate the specimen at a constant rate, and to continuously and simultaneously measure the instantaneous applied load (with a load cell) and the resulting elongations (using an extensometer). A stress-strain test typically takes several minutes to perform and is destructive; that is, the test specimen is permanently deformed and usually fractured.

Chapter 8 Mechanical Properties of Materials

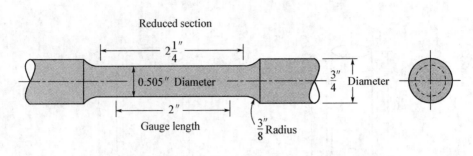

Figure 8.2 A standard tensile specimen with circular cross section

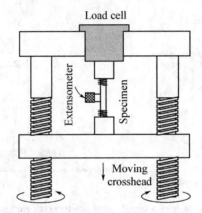

Figure 8.3 Schematic representation of the apparatus used to conduct tensile stress-train tests. The specimen is elongated by the moving crosshead; load cell and extensometer measure, respectively, the elongation

The output of such a tensile test is recorded. on a strip chart (or by a computer) as load or force versus elongation. These load-deformation[3] characteristics are dependent on the specimen size. For example, it will require twice the load to produce the same elongation if the cross-sectional area of the specimen is doubled. To minimize these geometrical factors, load and elongation are normalized to the respective parameters of engineering stress and engineering strain. Engineering stress σ is defined by the relationship

$$\sigma = \frac{F}{A_0} \tag{8-1}$$

in which F is the instantaneous load applied perpendicular to the specimen cross section, in units of newtons (N) or pounds force (lb_f), and A_0 is the original cross sectional area before any load is applied (m^2). The units of engineering stress (referred to subsequently as just stress) are megapascals, MPa (SI) (where 1 MPa = 10^6 N/m^2), and pounds force per square inch, psi (Customary U.S.).

Engineering strain ε is defined according to

$$\varepsilon = \frac{l_i - l_0}{l_0} = \frac{\Delta l}{l_0} \tag{8-2}$$

in which I_0 is the original length before any load is applied, and I_i is the instantaneous length. Sometimes the quantity $I_i - I_0$ is denoted as ΔI, and is the deformation elongation or change in length at some instant, as referenced to the original length. Engineering strain (subsequently called just strain) is unitless, but meters per meter or inches per inch are often used; the value of strain is obviously independent of the unit system. Sometimes strain is also expressed as a percentage, in which the strain value is multiplied by 100.

8.2.2 Compression Tests

Compression stress-strain tests may be conducted if in-service forces are of this type. A compression test is conducted in a manner similar to the tensile test, except that the force is compressive and the specimen contracts along the direction of the stress. Equations 8.1 and 8.2 are utilized to compute compressive stress and strain, respectively. By convention, a compressive force is taken to be negative, which yields a negative stress. Furthermore, since I_0 is greater than I_i, compressive strains computed from Equation 8.2 are necessarily also negative. Tensile tests are more common because they are easier to perform; also, for most materials used in structural applications, very little additional information is obtained from compressive tests. Compressive tests are used when a material's behavior under large and permanent (i.e., plastic) strains is desired, as in manufacturing applications, or when the material is brittle in tension.

8.2.3 Shear and Torsional Tests

For tests performed using a pure shear force as shown in Figure 8.1c, the shear stress τ is computed according to

$$\tau = \frac{F}{A_0} \quad (8-3)$$

where F is the load or force imposed parallel to the upper and lower faces, each of which has an area of A_0. The shear strain γ is defined as the tangent of the strain angle θ, as indicated in the figure. The units for shear stress and strain are the same as for their tensile counterparts.

Torsion is a variation of pure shear, wherein a structural member is twisted in the manner of Figure 8.1d; torsional forces produce a rotational motion about the longitudinal axis of one end of the member relative to the other end. Examples of torsion are found for machine axles and drive shafts, and also for twist drills.

Chapter 8　Mechanical Properties of Materials

Torsional tests are normally performed on cylindrical solid shafts or tubes. A shear stress τ is a function of the applied torque T, whereas shear strain γ is related to the angle of twist, ϕ in Figure 8.1d.

8.2.4　Geometric Considerations of the Stress State

Stresses that are computed from the tensile, compressive, shear and torsional force states represented in Figure 8.1 act either parallel or perpendicular to planar faces of the bodies represented in these illustrations. It should be noted that the stress state is a function of the orientations of the planes upon which the stresses are taken to act. For example, consider the cylindrical tensile specimen of Figure 8.4 that is subjected to a tensile stress σ applied parallel to its axis. Furthermore, consider also the plane p-p' that is oriented at some arbitrary angle θ relative to the plane of the specimen end-face[④]. Upon this plane p-p', the applied stress is no longer a pure tensile one. Rather, a more complex stress state is present that consists of a tensile (or normal) stress σ' that acts normal to the p-p' plane, and, in addition, a shear stress τ' that acts parallel to this plane; both of these stresses are represented in the figure. Using mechanics of materials principles, it is possible to develop equations for σ' and τ' in terms of σ and θ, as follows:

$$\sigma' = \sigma\cos^2\theta = \sigma\left(\frac{1+\cos 2\theta}{2}\right) \tag{8-4a}$$

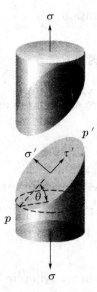

Figure 8.4　Schematic representation showing normal (σ') and shear (τ') stresses that act on a plane oriented at an angle θ relative to the plane taken perpendicular to the direction along which a pure tensile stress (σ) is applied

Chapter 8 Mechanical Properties of Materials

$$\tau' = \sigma \sin\theta \cos\theta = \sigma(\frac{\sin 2\theta}{2}) \qquad (8-4b)$$

These same mechanics principles allow the transformation of stress components from one coordinate system to another coordinate system that has a different orientation. Such treatments are beyond the scope of the present discussion.

8.3 Elastic Deformation

8.3.1 Stress-Strain Behavior

The degree to which a structure deforms or strains depends on the magnitude of an imposed stress. For most metals that are stressed in tension and at relatively low levels, stress and strain are proportional to each other through the relationship

$$\sigma = E\varepsilon \qquad (8-5)$$

This is known as Hooke's law, and the constant of proportionality E (GPa) is the modulus of elasticity or Young's modulus. For most typical metals the magnitude of this modulus ranges between 45 GPa, for magnesium, and 407 GPa, for tungsten. The moduli of elasticity are slightly higher for ceramic materials, which range between about 70 and 500 GPa. Polymers have modulus values that are smaller than both metals and ceramics, and which lie in the range 0.007 and 4 GPa. Room temperature modulus of elasticity values for a number of metals, ceramics and polymers are presented in Table 8.1.

Deformation in which stress and strain are proportional is called elastic deformation; a plot of stress (ordinate) versus strain (abscissa) results in a linear relationship, as shown in Figure 8.5. The slope of this linear segment corresponds to the modulus of elasticity E. This modulus may be thought of as stiffness, or a material's resistance to elastic deformation. The greater the modulus, the stiffer the material, or the smaller the elastic strain that results from the application of a given stress. The modulus is an important design parameter used for computing elastic deflections.

Elastic deformation is nonpermanent, which means that when the applied load is released, the piece returns to its original shape. As shown in the stress-strain plot (Figure 8.5), application of the load corresponds to moving from the origin up and along the straight line. Upon release of the load, the line is traversed in the opposite direction, back to the origin.

Chapter 8 Mechanical Properties of Materials

Table 8.1 Room-temperature elastic and shear moduli, and poisson's ratio for various materials

Material	Modulus of elasticity GPa	Shear modulus GPa	Poisson's ratio
Metal Alloys			
Tungsten	407	160	0.28
Steel	207	83	0.30
Nickel	207	76	0.31
Titanium	107	45	0.34
Copper	110	46	0.34
Brass	97	37	0.34
Aluminum	69	25	0.33
Magnesium	45	17	0.35
Ceramic Materials			
Aluminumoxide (Al_2O_3)	393	—	0.22
Siliconcarbide (SiC)	345	—	0.17
Siliconnitride (Si_3N_4)	304	—	0.30
Spinel ($MgAl_2O_4$)	260	—	—
Magnesiumoxide (MgO)	25	—	0.18
Zirconia	205	—	0.31
Mullite ($3Al_2O_3$—$2SiO_2$)	145	—	0.24
Glass-ceramic (Pyroceram)	120	—	0.25
Fusedsilica (SiO_2)	73	—	0.17
Soda-lime glass	69	—	0.23
Polymers			
Phenol-formaldehyde	2.76~4.83	—	—
Polyvinyl chloride (PVC)	2.41~4.14	—	0.38
Polyester (PET)	2.76~4.14	—	—
Polystyrene (PS)	2.28~3.28	—	0.33
Polymethyl methacrylate (PMMA)	2.24~3.24	—	—
Polycarbonate (PC)	2.38	—	0.36
Nylon 6, 6	1.58~3.80	—	0.39
Polypropylene (PP)	1.14~1.55	—	—
Polyethylene-high density (HDPE)	1.08	—	—
Polytetrafluoroethylene (PTFE)	0.40~0.55	—	0.46
Polyethylene-lowdensity (LDPE)	0.17~0.28	—	—

Chapter 8 Mechanical Properties of Materials

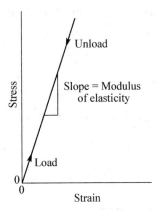

Figure 8.5 Schematic stress-strain diagram showing linear elastic deformation for loading and unloading cycles

There are some materials (e. g., gray cast iron, concrete and many polymers) for which this initial elastic portion of the stress-strain curve is not linear (Figure 8.6); hence, it is not possible to determine a modulus of elasticity as described above. For this nonlinear behavior, either tangent or secant modulus is normally used. Tangent modulus is taken as the slope of the stress-strain curve at some specified level of stress, while secant modulus represents the slope of a secant drawn from the origin to some given point of the $\sigma - \varepsilon$ curve. The determination of these moduli is illustrated in Figure 8.6.

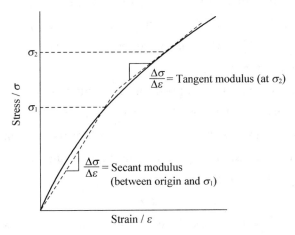

Figure 8.6 Schematic stress-strain diagram showing nonlinear elastic behavior and how secant and tangent moduli are determined

On an atomic scale, macroscopic elastic strain is manifested as small changes in the interatomic spacing and the stretching of interatomic bonds. As a consequence, the magnitude of the modulus of elasticity is a measure of the resistance to separation of adjacent atoms/ions/molecules, that is, the interatomic bonding forces. Furthermore, this modulus is proportional to the slope of the interatomic force-separation curve at the equilibrium spacing:

$$E \propto \left(\frac{dF}{dr}\right)_{r_0} \tag{8-6}$$

Figure 8.7 shows the force-separation curves for materials having both strong and weak interatomic bonds; the slope at r_0 is indicated for each.

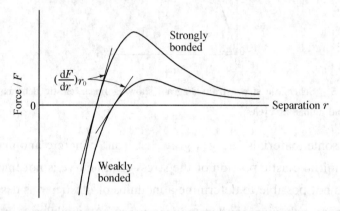

Figure 8.7 Force versus interatomic separation for weakly and strongly bonded atoms. The magnitude of the modulus of elasticity is proportional to the slope of each curve at the equilibrium interatomic separation r_0

Differences in modulus values between metals, ceramics and polymers are a direct consequence of the different types of atomic bonding that exist for the three materials types. Furthermore, with increasing temperature, the modulus of elasticity diminishes for all but some of the rubber materials; this effect is shown for several metals in Figure 8.8.

As would be expected, the imposition of compressive, shear or torsional stresses also evokes elastic behavior. The stress-strain characteristics at low stress levels are virtually the same for both tensile and compressive situations, to include the magnitude of the modulus of elasticity. Shear stress and strain are proportional to each other through the expression

$$\tau = G\gamma \tag{8-7}$$

where G is the shear modulus, the slope of the linear elastic region of the shear stress-strain curve. Table 8.1 also gives the shear moduli for a number of the common metals.

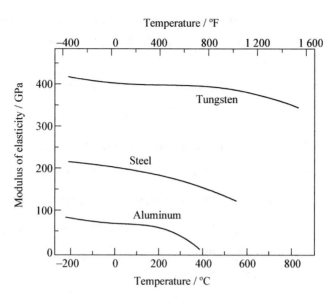

Figure 8.8 Plot of modulus of elasticity versus temperature for tungsten, steel and aluminum

8.3.2 Anelasticity

Up to this point, it has been assumed that elastic deformation is time independent, that is, that an applied stress produces an instantaneous elastic strain that remains constant over the period of time the stress is maintained. It has also been assumed that upon release of the load the strain is totally recovered, that is, that the strain immediately returns to zero. In most engineering materials, however, there will also exist a time-dependent elastic strain component. That is, elastic deformation will continue after the stress application, and upon load release some finite time is required for complete recovery. This time-dependent elastic behavior is known as anelasticity, and it is due to time-dependent microscopic and atomistic processes that are attendant to the deformation. For metals the anelastic component is normally small and is often neglected. However, for some polymeric materials its magnitude is significant; in this case it is termed viscoelastic behavior.

8.3.3 Elastic Properties of Materials

When a tensile stress is imposed on virtually all materials, an elastic elongation and accompanying strain ε_z result in the direction of the applied stress (arbitrarily taken to be the z direction), as indicated in Figure 8.9. As a result of this elongation, there will be constrictions in the lateral (x and y) directions perpendicular to the applied stress; from these contractions, the compressive strains ε_x and ε_y may be determined. If the applied stress is uniaxial (only in the z direction), and the material is isotropic, then $\varepsilon_x = \varepsilon_y$. A parameter termed Poisson's ratio v) is defined as the ratio of the lateral and axial strains, or

$$v = -\frac{\varepsilon_x}{\varepsilon_z} = -\frac{\varepsilon_y}{\varepsilon_z} \qquad (8-8)$$

The negative sign is included in the expression so that v will always be positive, since ε_x and ε_z will always be of opposite sign. Theoretically, Poisson's ratio for isotropic materials should be $\frac{1}{4}$; furthermore, the maximum value for v (or that value for which there is no net volume change) is 0.50. For many metals and other alloys, values of Poisson's ratio range between 0.25 and 0.35. Table 8.1 shows v values for several common materials.

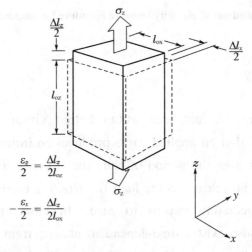

Figure 8.9 Axial (z) elongation (positive strain) and lateral (x and y) contractions (negative strains) in response to an imposed tensile stress. Solid lines represent dimensions after stress application; dashed lines, before

For isotropic materials, shear and elastic moduli are related to each other and to Poisson's ratio according to

$$E = 2G(1+v) \qquad (8-9)$$

In most metals G is about $0.4 E$; thus, if the value of one modulus is known, the other may be approximated.

Many materials are elastically anisotropic; that is, the elastic behavior (e.g., the magnitude of E) varies with crystallographic direction. For these materials the elastic properties are completely characterized only by the specification of several elastic constants, their number depending on characteristics of the crystal structure. Even for isotropic materials, for complete characterization of the elastic properties, at least two constants must be given. Since the grain orientation is random in most polycrystalline materials, these may be considered to be isotropic; inorganic ceramic glasses are also isotropic. The remaining discussion of mechanical behavior assumes isotropy and polycrystallinity (for metals and crystalline ceramics) because such is the character of most engineering materials.

8.4 Mechanical Behavior of Metals

For most metallic materials, elastic deformation persists only to strains of about 0.005. As the material is deformed beyond this point, the stress is no longer proportional to strain (Hooke's law[⑤], Equation 8.5, ceases to be valid), and permanent, nonrecoverable or plastic deformation occurs. Figure 8.10a plots schematically the tensile stress-strain behavior into the plastic region for a typical metal. The transition from elastic to plastic is a gradual one for most metals; some curvature results at the onset of plastic deformation, which increases more rapidly with rising stress.

From an atomic perspective, plastic deformation corresponds to the breaking of bonds with original atom neighbors and then reforming bonds with new neighbors as large numbers of atoms or molecules move relative to one another; upon removal of the stress they do not return to their original positions. This permanent deformation for metals is accomplished by means of a process called slip.

8.4.1 Tensile Properties

8.4.1.1 Yielding and Yield Strength

Most structures are designed to ensure that only elastic deformation will result when a stress is applied. It is therefore desirable to know the stress level at which plastic deformation begins, or where the phenomenon of yielding occurs. For metals that experience this gradual elastic-plastic transition, the point of yielding may be determined as the initial departure from linearity of the stress-strain curve; this is sometimes called the proportional limit, as indicated by point P in Figure 8.10a. In such cases the position of this point may not be determined precisely. As a consequence, a convention has been established wherein a straight line is constructed parallel to the elastic portion of the stress-strain curve at some specified strain offset, usually 0.002. The stress corresponding to the intersection of this line and the stress-strain curve as it bends over in the plastic region is defined as the yield strength σ_y. This is demonstrated in Figure 8.10a. Of course, the units of yield strength are MPa.

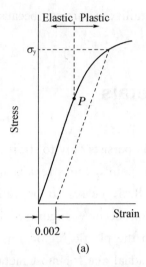 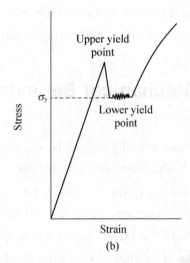

Figure 8.10 Typical stress-strain behavior for a metal showing elastic and plastic deformations, the proportional limit P, and the yield strength σ_y, as determined using the 0.002 strain offset method (a). Representative stress-strain behavior found for some steels demonstrating the yield point phenomenon (b)

For those materials having a nonlinear elastic region (Figure 8.6), use of the strain offset method is not possible, and the usual practice is to define the yield strength as the stress required to produce some amount of strain (e.g., $\varepsilon = 0.005$).

Some steels and other materials exhibit the tensile stress-strain behavior as shown in Figure 8.10b. The elastic-plastic transition is very well defined and occurs abruptly in what is termed a yield point phenomenon. At the upper yield point, plastic deformation is initiated with an actual decrease in stress. Continued deformation fluctuates slightly about some constant stress value, termed the lower yield point; stress subsequently rises with increasing strain. For metals that display this effect, the yield strength is taken as the average stress that is associated with the lower yield point, since it is well defined and relatively insensitive to the testing procedure. Thus, it is not necessary to employ the strain offset method for these materials.

The magnitude of the yield strength for a metal is a measure of its resistance to plastic deformation. Yield strengths may range from 35 MPa for low-strength aluminum to over 1 400 MPa for high-strength steels.

8.4.1.2 Tensile Strength

After yielding, the stress necessary to continue plastic deformation in metals increases to a maximum, point M in Figure 8.11, and then decreases to the eventual fracture, point F. The tensile strength TS (MPa) is the stress at the maximum on the engineering stress-strain curve (Figure 8.11). This corresponds to the maximum stress that can be sustained by a structure in tension; if this stress is applied and maintained, fracture will result. All deformation up to this point is uniform throughout the narrow region of the tensile specimen. However, at this maximum stress, a small constriction or neck begins to form at some point, and all subsequent deformation is confined at this neck, as indicated by the schematic specimen insets in Figure 8.11. This phenomenon is termed "necking", and fracture ultimately occurs at the neck. The fracture strength corresponds to the stress at fracture.

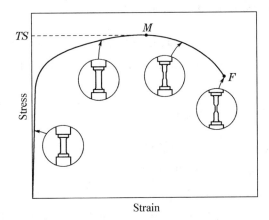

Figure 8.11 Typical engineering stress-strain behavior to fracture, point F. The tensile strength TS is indicated at point M. The circular insets represent the geometry of the deformed specimen at various points along the curve

Tensile strengths may vary anywhere from 50 MPa for an aluminum to as high as 3 000 MPa for the high-strength steels. Ordinarily, when the strength of a metal is cited for design purposes, the yield strength is used. This is because by the time a stress corresponding to the tensile strength has been applied, often a structure has experienced so much plastic deformation that it is useless. Furthermore, fracture strengths are not normally specified for engineering design purposes.

8.4.1.3 Ductility

Ductility is another important mechanical property. It is a measure of the degree of plastic deformation that has been sustained at fracture. A material that experiences very little or no plastic deformation upon fracture is termed brittle. The tensile stress-strain behaviors for both ductile and brittle materials are schematically illustrated in Figure 8.12.

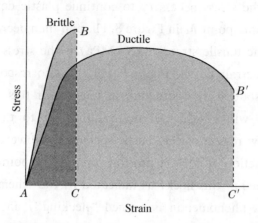

Figure 8.12 Schematic representations of tensile stress-strain behavior for brittle and ductile materials loaded to fracture

Ductility may be expressed quantitatively as either percent elongation or percent reduction in area. The percent elongation % EL is the percentage of plastic strain at fracture, or

$$\% \text{ EL} = \left(\frac{I_f - I_0}{I_0}\right) \times 100 \qquad (8-10)$$

where I_f is the fracture length and I_0 is the original gauge length as above. Inasmuch as a significant proportion of the plastic deformation at fracture is confined to the neck region, the magnitude of % EL will depend on specimen gauge length. The shorter I_0, the greater is the fraction of total elongation from the neck and, consequently, the higher the value of % EL. Therefore, I_0 should be specified when percent elongation values are cited; it is commonly 50 mm.

Percent reduction in area % RA is defined as

$$\% \text{ RA} = \left(\frac{A_0 - A_f}{A_0}\right) \times 100 \qquad (8-11)$$

where A_0 is the original cross-sectional area and A_f is the cross-sectional area at the point of fracture. Percent reduction in area values are independent of both I_0 and A_0. Furthermore, for a given material the magnitudes of % EL and % RA will, in general, be different. Most metals possess at least a moderate degree of ductility at room temperature; however, some become brittle as the temperature is lowered.

A knowledge of the ductility of materials is important for at least two reasons. First, it indicates to a designer the degree to which a structure will deform plastically before fracture. Second, it specifies the degree of allowable deformation during fabrication operations. We sometimes refer to relatively ductile materials as

Chapter 8 Mechanical Properties of Materials

being "forgiving", in the sense that they may experience local deformation without fracture should there be an error in the magnitude of the design stress calculation.

Brittle materials are approximately considered to be those having a fracture strain of less than about 5%.

Thus, several important mechanical properties of metals may be determined from tensile stress-strain tests. Table 8.2 presents some typical room-temperature values of yield strength, tensile strength and ductility for several common metals (and also for a number of polymers and ceramics). These properties are sensitive to any prior deformation, the presence of impurities, and/or any heat treatment to which the metal has been subjected. The modulus of elasticity is one mechanical parameter that is insensitive to these treatments. As with modulus of elasticity, the magnitudes of both yield and tensile strengths decline with increasing temperature; just the reverse holds for ductility, it usually increases with temperature. Figure 8.13 shows how the stress-strain behavior of iron varies with temperature.

Table 8.2 Room-temperature mechanical properties (in Tension) for various materials

Material	Yield strength MPa	Tensile strength MPa	Ductility, % EL [in 50 mm]
Metalalloys			
Molybdenum	565	655	35
Titanium	450	520	25
Steel (1020)	180	380	25
Nickel	138	480	40
Iron	130	262	45
Brass (70 Cu—30 Zn)	75	300	68
Copper	69	200	45
Aluminum	35	90	40
Ceramicmaterials			
Zirconia (ZrO_2)	—	800~1 500	—
Silicon nitride (Si_3N_4)	—	250~1 000	—
Aluminum oxide (Al_2O_3)	—	275~700	—
Silicon carbide (SiC)	—	100~820	—
Glass-ceramic (Pyroceram)	—	247	—
Mullite ($3Al_2O_3$—$2SiO_2$)	—	185	—
Spinel ($MgAl_2O_4$)	—	110~245	—
Fused silica (SiO_2)	—	110	—
Magnesium oxide (MgO)	—	105	—

Chapter 8　Mechanical Properties of Materials

Material	Yield strength MPa	Tensile strength MPa	Ductility, % EL [in 50 mm]
			(to be continued)
Soda-lime glass	—	69	—
Polymers			
Nylon 6,6	44.8~82.8	75.9~94.5	15~300
Polycarbonate (PC)	62.1	62.8~72.4	110~150
Polyester (PET)	59.3	48.3~72.4	30~300
Polymethyl methacrylate (PMMA)	53.8~73.1	48.3~72.4	2.0~5.5
Polyvinyl chloride (PVC)	40.7~44.8	40.7~51.7	40~80
Phenol-formaldehyde	—	34.5~62.1	1.5~2.0
Polystyrene (PS)	—	35.9~51.7	1.2~2.5
Polypropylene (PP)	31.0~37.2	31.0~41.4	100~600
Polyethylene-high density (HDPE)	26.2~33.1	22.1~31.0	10~1 200
Polytetrafluoroethylene (PTFE)	—	20.7~34.5	200~400
Polyethylene-low density (LDPE)	9.0~14.5	8.3~314	100~650

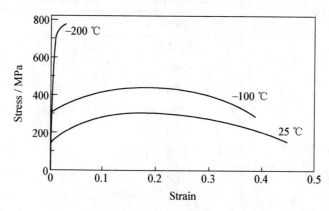

Figure 8.13　Engineering stress-strain behavior for iron at three temperatures

8.4.1.4　Resilience

Resilience is the capacity of a material to absorb energy when it is deformed elastically and then, upon unloading, to have this energy recovered. The associated property is the modulus of resilience, U_r, which is the strain energy per unit volume required to stress a material from an unloaded state up to the point of yielding.

Computationally, the modulus of resilience for a specimen subjected to a uniaxial tension test is just the area under the engineering stress-strain curve taken to yielding (Figure 8.14), or

$$U_r = \int_0^{\varepsilon_y} \sigma d\varepsilon \qquad (8-12a)$$

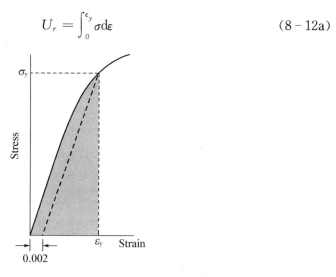

Figure 8.14 Schematic representation showing how modulus of resilience (corresponding to the shaded area) is determined from the tensile stress-strain behavior of a material

Assuming a linear elastic region,

$$U_r = \frac{1}{2}\sigma_y \varepsilon_y \qquad (8-12b)$$

in which ε_y is the strain at yielding.

The units of resilience are the product of the units from each of the two axes of the stress-strain plot. For SI units, this is joules per cubic meter (J/m^3, equivalent to Pa), whereas with Customary U.S. units it is inch-pounds force per cubic inch (in. lb_f/in^3). Both joules and inch-pounds force are units of energy, and thus this area under the stress-strain curve represents energy absorption per unit volume (in cubic meters or cubic inches) of material.

Incorporation of Equation 8.5 into Equation 8.12b yields

$$U_r = \frac{1}{2}\sigma_y \varepsilon_y = \frac{1}{2}\sigma_y \left(\frac{\sigma_y}{E}\right) = \frac{\sigma_y^2}{2E} \qquad (8-13)$$

Thus, resilient materials are those having high yield strengths and low moduli of elasticity; such alloys would be used in spring applications.

8.4.1.5 Toughness

Toughness is a mechanical term that is used in several contexts; loosely speaking, it is a measure of the ability of a material to absorb energy up to fracture. Specimen geometry as well as the manner of load application are important in toughness determinations. For dynamic (high strain rate) loading conditions and when a notch (or point of stress concentration) is present, notch toughness is assessed by using an impact test. Furthermore, fracture toughness is a property indicative of a material's resistance to fracture when a crack is present.

For the static (low strain rate) situation, toughness may be ascertained from the results of a tensile stress-strain test. It is the area under the $\sigma - \varepsilon$ curve up to the point of fracture. The units for toughness are the same as for resilience (i. e., energy per unit volume of material). For a material to be tough, it must display both strength and ductility; and often, ductile materials are tougher than brittle ones. This is demonstrated in Figure 8.12, in which the stress-strain curves are plotted for both material types. Hence, even though the brittle material has higher yield and tensile strengths, it has a lower toughness than the ductile one, by virtue of lack of ductility; this is deduced by comparing the areas ABC and $AB'C'$ in Figure 8.12.

8.4.2 True Stress and Strain

From Figure 8.11, the decline in the stress necessary to continue deformation past the maximum, point M. seems to indicate that the metal is becoming weaker. This is not at all the case; as a matter of fact, it is increasing in strength. However, the cross-sectional area is decreasing rapidly within the neck region, where deformation is occurring. This results in a reduction in the load-bearing capacity of the specimen. The stress, as computed from Equation 8.1, is on the basis of the original cross-sectional area before any deformation, and does not take into account this diminution in area at the neck.

Sometimes it is more meaningful to use a true stress-true strain scheme. True stress σ_T is defined as the load F divided by the instantaneous cross-sectional area A_i over which deformation is occurring (i. e., the neck, past the tensile point), or

$$\sigma_T = \frac{F}{A_i} \quad (8-14)$$

Furthermore, it is occasionally more convenient to represent strain as true strain ε_T, defined by

$$\varepsilon_T = \ln \frac{l_i}{l_0} \quad (8-15)$$

If no volume change occurs during deformation, that is, if

$$A_i l_i = A_0 l_0 \quad (8-16)$$

true and engineering stress and strain are related according to

$$\sigma_T = \sigma(1+\varepsilon) \quad (8-17a)$$

$$\varepsilon_T = \ln(1+\varepsilon) \quad (8-17b)$$

Equations 8.17a and 8.17b are valid only to the onset of necking; beyond this point true stress and strain should be computed from actual load, cross-sectional area and gauge length measurements.

A schematic comparison of engineering and true stress-strain behavior is made in Figure 8.15. It is worth noting that the true stress necessary to sustain increasing strain continues to rise past the tensile point M'.

Coincident with the formation of a neck is the introduction of a complex stress state within the neck region (*i.e.*, the existence of other stress components in addition to the axial stress). As a consequence, the correct stress (axial) within the neck is slightly lower than the stress computed from the applied load and neck cross-sectional area. This leads to the "corrected" curve in Figure 8.15.

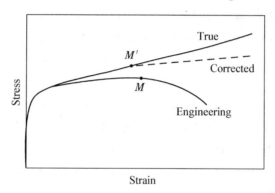

Figure 8.15 A comparison of typical tensile engineering stress-strain and true stress-strain behaviors. Necking begins at point M on the engineering curve, which corresponds to M' on the true curve. The "corrected" true stress-strain curve takes into account the complex stress state within the neck region

For some metals and alloys the region of the true stress-strain curve from the onset of plastic deformation to the point at which necking begins may be approximated by

$$\sigma_T = K\varepsilon_T^n \qquad (8-18)$$

Table 8.3 Tabulation of n and K values (Equation 8.18) for several alloys

Material	n	K / MPa
Low-carbon steel (annealed)	0.26	530
Alloy steel (Type 4340, annealed)	0.15	640
Stainless steel (Type 304, annealed)	0.45	1 275
Aluminum (annealed)	0.20	180
Aluminum alloy (Type 2024, heat treated)	0.16	690
Copper (annealed)	0.54	315
Brass (70Cu—30Zn, annealed)	0.49	895

In this expression, K and n are constants, which values will vary from alloy to alloy, and will also depend on the condition of the material (*i.e.*, whether it has been plastically deformed, heat treated, *etc.*). The parameter n is often termed the strain-hardening exponent and has a value less than unity. Values of n and K for several alloys are contained in Table 8.3.

8.4.3 Elastic Recovery during Plastic Deformation

Upon release of the load during the course of a stress-strain test, some fraction of the total deformation is recovered as elastic strain. This behavior is demonstrated in Figure 8.16, a schematic engineering stress-strain plot. During the unloading cycle, the curve traces a near straight-line path from the point of unloading (point D), and its slope is virtually identical to the modulus of elasticity, or parallel to the initial elastic portion of the curve. The magnitude of this elastic strain, which is regained during unloading, corresponds to the strain recovery, as shown in Figure 8.16. If the load is reapplied, the curve will traverse essentially the same linear portion in the direction opposite to unloading; yielding will again occur at the unloading stress level where the unloading began. There will also be an elastic strain recovery associated with fracture.

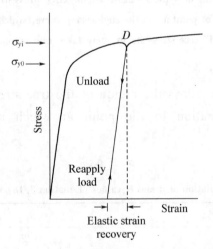

Figure 8.16 Schematic tensile stress-strain diagram showing the phenomena of elastic strain recovery and strain hardening. The initial yield strength is designated as σ_{y_0}; σ_{y_i} is the yield strength after releasing the load at point D, and then upon reloading

8.4.4 Compressive, Shear and Torsional Deformation

Of course, metals may experience plastic deformation under the influence of applied compressive, shear and torsional loads. The resulting stress-strain behavior into the plastic region will be similar to the tensile counterpart (Figure 8.10a: yielding and the associated curvature). However, for compression, there will be no maximum, since necking does not occur; furthermore, the mode of fracture will be different from that for tension.

8.5 Mechanical Behavior-Ceramics

Ceramic materials are somewhat limited in applicability by their mechanical properties, which in many respects are inferior to those of metals. The principal drawback is a disposition to catastrophic fracture in a brittle manner with very little energy absorption. In this section we explore the salient mechanical characteristics of these materials and how these properties are measured.

8.5.1 Flexural Strength

The stress-strain behavior of brittle ceramics is not usually ascertained by a tensile test as outlined in Section 8.2, for three reasons. First, it is difficult to prepare and test specimens having the required geometry. Second, it is difficult to grip brittle materials without fracturing them. Third, ceramics fail after only about 0.1% strain, which necessitates that tensile specimens be perfectly aligned in order to avoid the presence of bending stresses, which are not easily calculated. Therefore, a more suitable transverse bending test is most frequently employed, in which a rod specimen having either a circular or rectangular cross section is bent until fracture using a three- or four-point loading technique: the three-point loading scheme is illustrated in Figure 8.17. At the point of loading, the top surface of the specimen is placed in a state of compression, whereas the bottom surface is in tension. Stress is computed from the specimen thickness, the bending moment, and the moment of inertia of the cross section; these parameters are noted in Figure 8.17 for rectangular and circular cross sections. The maximum tensile stress (as determined using these stress expressions) exists at the bottom specimen surface directly below the point of load application. Since the tensile strengths of ceramics are about one-tenth of their compressive strengths, and since fracture occurs on the tensile specimen face, the flexure test is a reasonable substitute for the tensile test.

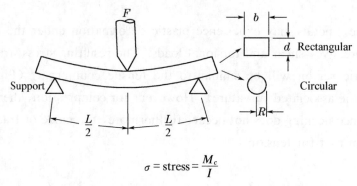

$$\sigma = \text{stress} = \frac{M c}{I}$$

where M = maximum bending moment
c = distance from center of specimen to outer fibers
I = moment of interia of cross section
F = applied load

	M'	c	I	σ
Rectangular	$\frac{FL}{4}$	$\frac{d}{2}$	$\frac{bd^3}{12}$	$\frac{3FL}{2bd^2}$
Circular	$\frac{FL}{4}$	R	$\frac{\pi R^4}{4}$	$\frac{FL}{\pi R^3}$

Figure 8.17 A three-point loading scheme for measuring the stress-strain behavior and flexural strength of brittle ceramics, including expressions for computing stress for rectangular and circular cross sections

The stress at fracture using this flexure test is known as the flexural strength, modulus of rupture, fracture strength, or the bend strength, an important mechanical parameter for brittle ceramics. For a rectangular cross section, the flexural strength σ_{fs} is equal to

$$\sigma_{fs} = \frac{3F_f L}{2bd^2} \tag{8-19a}$$

where F_f is the load at fracture, L is the distance between support points, and the other parameters are as indicated in Figure 8.17. When the cross section is circular, then

$$\sigma_{fs} = \frac{F_f L}{\pi R^3} \tag{8-19b}$$

R being the specimen radius.

Characteristic flexural strength values for several ceramic materials are given in Table 8.2. Since, during bending, a specimen is subjected to both compressive and tensile stresses, the magnitude of its flexural strength is greater than the tensile fracture strength. Furthermore, σ_{fs} will depend on specimen size with increasing specimen volume (under stress) there is an increase in the probability of the existence of a crack-producing flaw, and consequently, a decrease in flexural strength.

8.5.2 Elastic Behavior

The elastic stress-strain behavior for ceramic materials using these flexure tests is similar to the tensile test results for metals: a linear relationship exists between stress and strain. Figure 8.18 compares the stress-strain behavior to fracture for aluminum oxide (alumina) and glass. Again, the slope in the elastic region is the modulus of elasticity; also, the moduli of elasticity for ceramic materials are slightly higher than for metals (Table 8.2). From Figure 8.18 it may be noted that neither of the materials experiences plastic deformation prior to fracture.

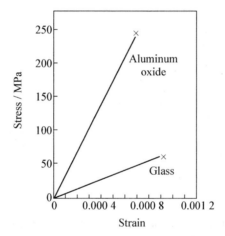

Figure 8.18   Typical stress-strain behavior to fracture for aluminum oxide and glass

8.6 Mechanical Behavior of Polymers

8.6.1 Stress-Strain Behavior

The mechanical properties of polymers are specified with many of the same parameters that are used for metals, that is, modulus of elasticity, and yield and tensile strengths. For many polymeric materials, the simple stress-strain test is employed for the characterization of some of these mechanical parameters. The mechanical characteristics of polymers, for the most part, are highly sensitive to the rate of deformation (strain rate), the temperature and the chemical nature of the

Chapter 8 Mechanical Properties of Materials

environment (the presence of water, oxygen, organic solvents, etc.). Some modifications of the testing techniques and specimen configurations used for metals are necessary with polymers, especially for the highly elastic materials, such as rubbers.

Three typically different types of stress-strain behavior are found for polymeric materials, as represented in Figure 8.19. Curve A illustrates the stress-strain character for a brittle polymer, inasmuch as it fractures while deforming elastically. The behavior for the plastic material, curve B, is similar to that found for many metallic materials; the initial deformation is elastic, which is followed by yielding and a region of plastic deformation. Finally, the deformation displayed by curve C is totally elastic; this rubber like elasticity (large recoverable strains produced at low stress levels) is displayed by a class of polymers termed the elastomers.

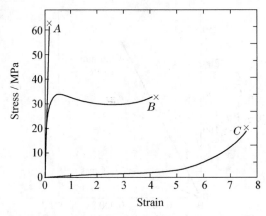

Figure 8.19 The stress-strain behavior for brittle (curve A), plastic (curve B) and highly elastic (elastomeric) (curve C) polymers

Modulus of elasticity (termed tensile modulus or sometimes just modulus for polymers) and ductility in percent elongation are determined for polymers in the same manner as for metals (Section 8.4.1). For plastic polymers (curve B, Figure 8.19), the yield point is taken as a maximum on the curve, which occurs just beyond the termination of the linear-elastic region (Figure 8.20); the stress at this maximum is the yield strength (σ_y). Furthermore, tensile strength (TS) corresponds to the stress at which fracture occurs (Figure 8.20); TS may be greater than or less than cry. Strength, for these plastic polymers, is normally taken as tensile strength. Table 8.2 gives these mechanical properties for a number of polymeric materials.

Chapter 8 Mechanical Properties of Materials

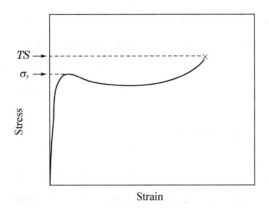

Figure 8.20 Schematic stress-strain curve for a plastic polymer showing how yield and tensile strengths are determined

Polymers are, in many respects, mechanically dissimilar to metals (and ceramic materials). For example, the modulus for highly elastic polymeric materials may be as low as 7 MPa, but may run as high as 4 GPa for some of the very stiff polymers; modulus values for metals are much larger (Table 8.1). Maximum tensile strengths for polymers are on the order of 100 MPa for some metal alloys 4100 MPa. And, whereas metals rarely elongate plastically to more than 100 %, some highly elastic polymers may experience elongations to as much as 1000 %.

In addition, the mechanical characteristics of polymers are much more sensitive to temperature changes within the vicinity of room temperature. Consider the stress-strain behavior for polymethyl methacrylate (Plexiglas) at several temperatures between 4 and 60 ℃ (40 and 140 ℉) (Figure 8.21). Several features of this figure are worth noting, as follows: increasing the temperature produces (1) a decrease in elastic modulus, (2) a reduction in tensile strength, and (3) an enhancement of ductility-at 4 ℃ (40 ℉) the material is totally brittle, whereas considerable plastic deformation is realized at both 50 and 60 ℃ (122 and 140 ℉).

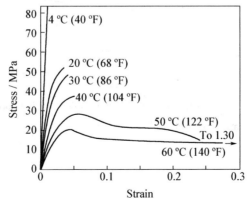

Figure 8.21 The influence of temperature on the stress-strain characteristics of polymethyl methacrylate

The influence of strain rate on the mechanical behavior may also be important. In general, decreasing the rate of deformation has the same influence on the stress-strain characteristics as increasing the temperature; that is, the material becomes softer and more ductile.

8.6.2 Macroscopic Deformation

Some aspects of the macroscopic deformation of semicrystalline polymers deserve our attention. The tensile stress-strain curve for a semicrystalline material, which was initially unoriented, is shown in Figure 8.22; also included in the figure are schematic representations of specimen profile at various stages of deformation. Both upper and lower yield points are evident on the curve, which are followed by a near horizontal region. At the upper yield point, a small neck forms within the gauge section of the specimen. Within this neck, the chains become oriented (i. e., chain axes become aligned parallel to the elongation direction, a condition that is represented schematically), which leads to localized strengthening. Consequently, there is a resistance to continued deformation at this point, and specimen elongation proceeds by the propagation of this neck region along the gauge length; the chain orientation phenomenon accompanies this neck extension. This tensile behavior may be contrasted to that found for ductile metals (Section 8.4.1), wherein once a neck has formed, all subsequent deformation is confined to within the neck region.

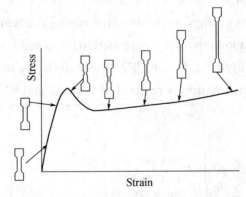

Figure 8.22 Schematic tensile stress-strain curve for a semicrystalline polymer. Specimen contours at several stages of deformation are included

Chapter 8 Mechanical Properties of Materials

8.7 Hardness and Other Mechanical Property Considerations

8.7.1 Hardness

Another mechanical property that may be important to consider is hardness, which is a measure of a material's resistance to localized plastic deformation (e. g., a small dent or a scratch). Early hardness tests were based on natural minerals with a scale constructed solely on the ability of one material to scratch another that was softer. A qualitative and somewhat arbitrary hardness indexing scheme was devised, termed the Mohs scale[⑥], which ranged from 1 on the soft end for talc to 10 for diamond. Quantitative hardness techniques have been developed over the years in which a small indenter is forced into the surface of a material to be tested, under controlled conditions of load and rate of application. The depth or size of the resulting indentation is measured, which in turn is related to a hardness number; the softer the material, the larger and deeper the indentation, and the lower the hardness index number. Measured hardnesses are only relative (rather than absolute), and care should be exercised when comparing values determined by different techniques.

Hardness tests are performed more frequently than any other mechanical test for several reasons:

(1) They are simple and inexpensive-ordinarily no special specimen need be prepared, and the testing apparatus is relatively inexpensive.

(2) The test is nondestructive-the specimen is neither fractured nor excessively deformed; a small indentation is the only deformation.

(3) Other mechanical properties often may be estimated from hardness data, such as tensile strength (see Figure 8.23).

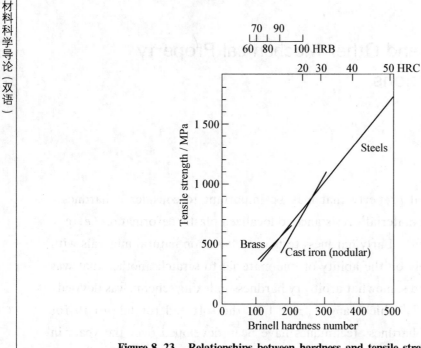

Figure 8.23　Relationships between hardness and tensile strength for steel, brass and cast iron

8.7.2　Rockwell Hardness Tests

The Rockwell tests constitute the most common method used to measure hardness because they are so simple to perform and require no special skills. Several different scales may be utilized from possible combinations of various indenters and different loads, which permit the testing of virtually all metal alloys (as well as some polymers). Indenters include spherical and hardened steel balls having diameters of $\frac{1}{16}$, $\frac{1}{8}$, $\frac{1}{4}$ and $\frac{1}{2}$ in. (1.588, 3.175, 6.350 and 12.70 mm), and a conical diamond (Brale) indenter, which is used for the hardest materials.

With this system, a hardness number is determined by the difference in depth of penetration resulting from the application of an initial minor load followed by a larger major load; utilization of a minor load enhances test accuracy. On the basis of the magnitude of both major and minor loads, there are two types of tests: Rockwell and superficial Rockwell. For Rockwell, the minor load is 10 kg, whereas major loads are 60, 100 and 150 kg. Each scale is represented by a letter of the alphabet; several are listed with the corresponding indenter and load in Tables 8.4 and 8.5a. For superficial tests, 3 kg is the minor load; 15, 30 and 45 kg are the possible major load values. These scales are identified by a 15, 30 or 45 (according

Table 8.4 Hardness testing techniques

Test	Indenter	Shape of indentation (Side view)	Shape of indentation (Top view)	Load	Formula for hardness number
Brinell	10-mm sphere of steel or tungsten carbide			P	$HB = \dfrac{2P}{\pi D[D - \sqrt{D^2 - d^2}]}$
Vickers microhardness	Diamond pyramid	136°	d_1	P	$HV = \dfrac{1.854 P}{d_1^2}$
Knoop microhardness	Diamond pyramid	$l/b = 7.11$, $b/t = 4.00$	b	P	$HK = \dfrac{14.2 P}{l^2}$
Rockwell and Superficial Rockwell	Diamond Cone $\dfrac{1}{16}$, $\dfrac{1}{8}$, $\dfrac{1}{4}$, $\dfrac{1}{2}$ in. Diameter steel spheres	120°		60 kg 100 kg } Rockwell 150 kg 15 kg 30 kg } Superficial Rockwell 45 kg	

to load), followed by N, T, W, X or Y, depending on indenter. Superficial tests are frequently performed on thin specimens. Table 8.5b presents several superficial scales.

When specifying Rockwell and superficial hardnesses, both hardness number and scale symbol must be indicated. The scale is designated by the symbol HR followed by the appropriate scale identification. For example, 80 HRB represents a Rockwell hardness of 80 on the B scale, and 60 HR30W indicates a superficial hardness of 60 on the 30 W scale.

For each scale, hardnesses may range up to 130; however, as hardness values rise above 100 or drop below 20 on any scale, they become inaccurate; and because the scales have some overlap, in such a situation it is best to utilize the next harder or softer scale.

Inaccuracies also result if the test specimen is too thin, if an indentation is made too near a specimen edge, or if two indentations are made too close to one another. Specimen thickness should be at least ten times the indentation depth, whereas allowance should be made for at least three indentation diameters between the center of one indentation and the specimen edge, or to the center of a second indentation. Furthermore, testing of specimens stacked one on top of another is not recommended. Also, accuracy is dependent on the indentation being made into a smooth flat surface.

Table 8.5a Rockwell hardness scales

Scalesymbol	Indenter	Majorload /kg
A	Diamond	60
B	$\frac{1}{16}$ in. ball	100
C	Diamond	150
D	Diamond	100
E	$\frac{1}{8}$ in. ball	100
F	$\frac{1}{16}$ in. ball	60
G	$\frac{1}{16}$ in. ball	150
H	$\frac{1}{8}$ in. ball	60
K	$\frac{1}{8}$ in. ball	150

Chapter 8 Mechanical Properties of Materials

Table 8.5b Superficial rockwell hardness scales

Scalesymbol	Indenter	Majorload /kg
15N	Diamond	15
30N	Diamond	30
45N	Diamond	45
15T	$\frac{1}{16}$ in. ball	15
30T	$\frac{1}{16}$ in. ball	30
45T	$\frac{1}{16}$ in. ball	45
15W	$\frac{1}{8}$ in. ball	15
30W	$\frac{1}{8}$ in. ball	30
45W	$\frac{1}{8}$ in. ball	45

The modern apparatus for making Rockwell hardness measurements is automated and very simple to use; hardness is read directly, and each measurement requires only a few seconds.

The modern testing apparatus also permits a variation in the time of load application. This variable must also be considered in interpreting hardness data.

8.7.3 Brinell Hardness Tests

In Brinell tests, as in Rockwell measurements, a hard, spherical indenter is forced into the surface of the metal to be tested. The diameter of the hardened steel (or tungsten carbide) indenter is 10.00 mm (0.394 in.). Standard loads range between 500 and 3 000 kg in 500 kg increments; during a test, the load is maintained constant for a specified time (between 10 and 30 s). Harder materials require greater applied loads. The Brinell hardness number, HB, is a function of both the magnitude of the load and the diameter of the resulting indentation (see Table 8.4). This diameter is measured with a special low-power microscope, utilizing a scale that is etched on the eyepiece. The measured diameter is then converted to the appropriate HB number using a chart; only one scale is employed with this technique.

Maximum specimen thickness as well as indentation position (relative to specimen edges) and minimum indentation spacing requirements are the same as for Rockwell tests. In addition, a well-defined indentation is required; this necessitates a smooth flat surface in which the indentation is made.

8.7.4 Knoop and Vickers Microhardness Tests

Two other hardness testing techniques are Knoop (pronounced nūp) and Vickers (sometimes also called diamond pyramid). For each test a very small diamond indenter having pyramidal geometry is forced into the surface of the specimen. Applied loads are much smaller than for Rockwell and Brinell, ranging between 1 and 1 000 g. The resulting impression is observed under a microscope and measured; this measurement is then converted into a hardness number (Table 8.4). Careful specimen surface preparation (grinding and polishing) may be necessary to ensure a well-defined indentation that may be accurately measured. The Knoop and Vickers hardness numbers are designated by HK and HV, respectively, and hardness scales for both techniques are approximately equivalent. Knoop and Vickers are referred to as microhardness testing methods on the basis of load and indenter size. Both are well suited for measuring the hardness of small, selected specimen regions; furthermore, Knoop is used for testing brittle materials such as ceramics.

There are other hardness-testing techniques that are frequently employed, but which will not be discussed here; these include ultrasonic microhardness, dynamic (Scleroscope), durometer (for plastic and elastomeric materials) and scratch hardness tests. These are described in references provided at the end of the chapter.

8.7.5 Hardness Conversion

The facility to convert the hardness measured on one scale to that of another is most desirable. However, since hardness is not a well-defined material property, and because of the experimental dissimilarities among the various techniques, a comprehensive conversion scheme has not been devised. Hardness conversion data have been determined experimentally and found to be dependent on material type and characteristics. Detailed conversion tables for various other metals and alloys are contained in ASTM Standard E 140, "Standard Hardness Conversion Tables for Metals". In light of the preceding discussion, care should be exercised in extrapolation of conversion data from one alloy system to another.

8.7.6 Correlation between Hardness and Tensile Strength

Both tensile strength and hardness are indicators of a metal's resistance to plastic deformation. Consequently, they are roughly proportional, as shown in Figure 8.23, for tensile strength as a function of the HB for cast iron, steel and brass. The same proportionality relationship does not hold for all metals, as Figure 8.23 indicates. As a rule of thumb for most steels, the HB and the tensile strength are related according to

$$TS(\text{MPa}) = 3.45 \times \text{HB} \qquad (8-20)$$

8.7.7 Hardness of Ceramic Materials

One beneficial mechanical property of ceramics is their hardness, which is often utilized when an abrasive or grinding action is required; in fact, the hardest known materials are ceramics. A listing of a number of different ceramic materials according to Knoop hardness is contained in Table 8.6. Only ceramics having Knoop hardnesses of about 1 000 or greater are utilized for their abrasive characteristics.

Table 8.6 Approximate Knoop hardness (100 g load) for seven ceramic materials

Material	Approximate Knoop hardness
Diamond (carbon)	7 000
Boron carbide (B_4C)	2 800
Silicon carbide (SiC)	2 500
Tungsten carbide (WC)	2 100
Aluminum oxide (Al_2O_3)	2 100
Quartz (SiO_2)	800
Glass	550

8.7.8 Tear Strength and Hardness of Polymers

Mechanical properties that are sometimes influential in the suitability of a polymer for some particular application include tear resistance and hardness. The ability to resist tearing is an important property of some plastics, especially those used for thin films in packaging. Tear strength, the mechanical parameter that is measured, is the energy required to tear apart a cut specimen that has a standard geometry. The magnitude of tensile and tear strengths are related.

Polymers are softer than metals and ceramics, and most hardness tests are conducted by penetration techniques similar to those described for metals in the previous section. Rockwell tests are frequently used for polymers. Other indentation techniques employed are the Durometer and Barcol.

8.8 Property Variability and Design/Safety Factors

8.8.1 Variability of Material Properties

At this point it is worthwhile to discuss an issue that sometimes proves troublesome to many engineering students, namely, that measured material properties are not exact quantities. That is, even if we have a most precise measuring apparatus and a highly controlled test procedure, there will always be some scatter or variability in the data that are collected from specimens of the same material. For example, consider a number of identical tensile samples that are prepared from a single bar of some metal alloy, which samples are subsequently stress-strain tested in the same apparatus. We would most likely observe that each resulting stress-strain plot is slightly different from the others. This would lead to a variety of modulus of elasticity, yield strength and tensile strength values. A number of factors lead to uncertainties in measured data. These include the test method, variations in specimen fabrication procedures, operator bias and apparatus calibration. Furthermore, inhomogeneities may exist within the same lot of material, and/or slight compositional and other differences from lot to lot. Of course, appropriate measures should be taken to minimize the possibility of measurement error, and also to mitigate those factors that lead to data variability.

It should also be mentioned that scatter exists for other measured material properties such as density, electrical conductivity and coefficient of thermal expansion.

It is important for the design engineer to realize that scatter and variability of materials properties are inevitable and must be dealt with appropriately. On occasion, data must be subjected to statistical treatments and probabilities determined. For example, instead of asking the question, "What is the fracture strength of this alloy"? The engineer should become accustomed to asking the question, "What is the probability of failure of this alloy under these given circumstances"?

It is often desirable to specify a typical value and degree of dispersion (or scatter) for some measured property; such is commonly accomplished by taking the average and the standard deviation, respectively.

8.8.2 Design/Safety Factors

There will always be uncertainties in characterizing the magnitude of applied loads and their associated stress levels for in-service applications; ordinarily load calculations are only approximate. Furthermore, as noted in the previous section, virtually all engineering materials exhibit a variability in their measured mechanical properties. Consequently, de sign allowances must be made to protect against unanticipated failure. One way this may be accomplished is by establishing, for the particular application, a design stress, denoted as σ_d. For static situations and when ductile materials are used, σ_d is taken as the calculated stress level σ_c (on the basis of the estimated maximum load) multiplied by a design factor, N', that is

$$\sigma_d = N' \sigma_c \qquad (8-20)$$

where N' is greater than unity. Thus, the material to be used for the particular application is chosen so as to have a yield strength at least as high as this value of σ_d.

Alternatively, a safe stress or working stress, σ_w, is used instead of design stress. This safe stress is based on the yield strength of the material and is defined as the yield strength divided by a factor of safety, N, or

$$\sigma_w = \frac{\sigma_y}{N} \qquad (8-21)$$

Utilization of design stress (Equation 8-20) is usually preferred since it is based on the anticipated maximum applied stress instead of the yield strength of the material; normally there is a greater uncertainty in estimating this stress level than in the specification of the yield strength. However, in the discussion of this text, we are concerned with factors that influence yield strengths, and not in the determination of applied stresses; therefore, the succeeding discussion will deal with working stresses and factors of safety.

The choice of an appropriate value of N is necessary. If N is too large, then component over design will result, that is, either too much material or a material having a higher-than-necessary strength will be used. Values normally range between 1.2 and 4.0. Selection of N will depend on a number of factors, including

Chapter 8 Mechanical Properties of Materials

economics, previous experience, the accuracy with which mechanical forces and material properties may be determined, and, most important, the consequences of failure in terms of loss of life and/or property damage.

◆ Notes

① well-defined 定义明确的；界限清楚的
② stress-strain 应力—应变
③ load-deformation 载荷变形
④ end-face 端面
⑤ Hooke's law 胡克定律
⑥ Mohs scale 莫氏硬度

◆ Vocabulary

abscissa [æbˈsɪsə] n. [数]横坐标；横线
alumina [əˈluːmɪnə] n. [无化]氧化铝；矾土
anelastic [ˌænɪˈlæstɪk] adj. 滞弹(性)的
anelasticity [ˌænɪəlæsˈtɪsətɪ] n. [物]滞弹性；[力]内摩擦力
brale [breɪl] n. [冶]金刚石压头
Brinell [ˈbrɪnel] n. 布氏硬度
catastrophic [ˌkætəˈstrɒfɪk] adj. 灾难的；悲惨的；灾难性的，毁灭性的
compressive [kəmˈpresɪv] adj. 压缩的；有压缩力的
conical [ˈkɒnɪkəl] adj. 圆锥的；圆锥形的
constriction [kənˈstrʌkʃən] 建造，结构，建筑
conversion [kənˈvɜːʃən] n. 转换；变换
deflection [dɪˈflekʃən] n. 偏向；挠曲；偏差
dent [dent] n. 凹痕；削弱；减少；齿 vt. 削弱；使产生凹痕 vi. 产生凹陷；凹进去；削减
drawback [ˈdrɔːbæk] n. 缺点，不利条件
ductility [dʌkˈtɪlətɪ] n. 延展性；柔软性
duration [djʊˈreɪʃən] n. 持续，持续的时间，期间

durometer [djʊəˈrɒmɪtə] n. 硬度计；硬度测验器
elastomer [iˈlæstəmə] n. [力]弹性体，[高分子]高弹体
elongate [ˈiːlɒŋɡeɪt] adj. 伸长的；延长的 vi. 拉长；延长；伸长 vt. 拉长；使延长；使伸长
elongation [ˌiːlɒŋˈɡeɪʃən] n. 伸长；伸长率；延伸率；延长
extensometer [ˌekstenˈsɒmɪtə] n. [力]伸长计，延伸仪
flexural [ˈflekʃərəl] adj. 弯曲的；曲折的
fluctuate [ˈflʌktʃueɪt] vi. 波动；涨落；动摇 vt. 使波动；使动摇
gauge [ɡeɪdʒ] n. 计量器；标准尺寸；容量规格 vt. 测量；估计；给……定规格
indenter [ɪnˈdentə] n. 硬度计压头
inertia [ɪˈnɜːʃə] n. [力]惯性
joules [dʒuːlz] n. 焦耳(joule 的复数)
knoop [nuːp] n. 努普(硬度单位)；努普显微压痕硬度试验仪
longitudinal [ˌlɒndʒɪˈtjuːdɪnəl] adj. 长度的，纵向的；经线的
methacrylate [mɪˈθækrɪleɪt] n. 甲基丙烯酸酯；异丁烯酸甲酯

Chapter 8 Mechanical Properties of Materials

nonpermanent [ˈnɒnpɜːmənənt] *adj.* 非永久的
nonrecoverable [ˌnɒnrɪˈkʌvərəbl] *adj.* 不能收回的;不可恢复的
notch [nɒtʃ] *n.* (V字形)槽口,凹口;(做记录、记数的)刻痕 *vt.* 用刻痕计算;在……上刻凹痕
ordinate [ˈɔːdɪnət] *n.* [数]纵坐标
poisson [pwaːsɒn] *n.* 泊松
propagation [ˌprɒpəˈɡeɪʃən] *n.* [物理学]传播,传送
pyramid [ˈpɪrəmɪd] *n.* 角锥体
resilience [rɪˈzɪlɪəns] *n.* 恢复力;弹力
rockwell [ˈrɒkwel] *n.* 铬氏硬度

strain [streɪn] *n.* 张力;拉紧
tangent [ˈtændʒənt] *n.* [数]切线,[数]正切 *adj.* 切线的,相切的
tensile [ˈtensaɪl] *adj.* [力]拉力的;可伸长的;可拉长的
torque [tɔːk] *n.* 转矩,[力]扭矩
torsion [ˈtɔːʃən] *n.* 扭转,扭曲;转矩,[力]扭力
torsional [ˈtɔːʃənəl] *adj.* 扭转的;扭力的
toughness [ˈtʌfnɪs] *n.* [力]韧性
uniaxial [ˌjuːnɪˈæksɪəl] *adj.* 单轴的
viscoelastic [ˌvɪskəʊɪˈlæstɪk] *adj.* [力]粘弹性的

Exercises

1. Translate the following Chinese phrases into English

(1) 应力分析   (2) 应力—应变测试   (3) 试样尺寸
(4) 作用扭矩   (5) 弹性形变   (6) 线性关系
(7) 工程材料   (8) 屈服强度   (9) 抗拉强度
(10) 高强度钢   (11) 低碳钢

2. Translate the following English phrases into Chinese

(1) mechanical behavior   (2) gauge length
(3) compression tests   (4) shear strain
(5) ceramic materials   (6) nonlinear behavior
(7) elastic properties   (8) proportional limit
(9) yield point   (10) mechanical parameter

3. Translate the following Chinese sentences into English

(1) 与此相反,力学性质可以被认为主要是原子间的相互作用。
(2) 碳的含量高于0.8%,结构中会出现游离的Fe_3C,抗拉强度开始降低。
(3) 最重要的事是如何提高材料的强度。
(4) 当硬材料的硬度值差别比较小的时候,洛氏硬度值就不像维氏硬度值那么准确。
(5) 当对材料施加载荷时,它会产生应力。
(6) 通常来说,屈服强度是最重要的性能,高于它,材料就会发生永久变形。

4. Translate the following English sentences into Chinese

(1) Soft metals creep at room temperature, but most metallic materials only show the effect at higher temperature.

(2) In assessing the behavior of a metal under stress, especially at elevatedtemperature, it is necessary to consider the time factor.

(3) Significant improvements to the fracture toughness, ductility and impact resistance of ceramics wererealized and thus the gap in physical properties between ceramics and metals began to close.

(4) Stress is defined as the force divided by the cross sectional area on which the force acts.

(5) In the Rockwell test the depth of the indentation is measured by the instrument and this is directly indicated on a dial as a hardness value.

5. Translate the following Chinese essay into English

随着晶粒尺寸的减小,材料的硬度增大。然而,具有小的拉伸应力样品的应力——应变曲线表明,其屈服应力几乎没什么变化。最近,一些综述对纳米结构材料中的硬度和屈服应力的力学行为给予了总结,证实了当晶粒尺寸减小至纳米范围(<100nm),材料的硬度随粒径的减小而增大。

6. Translate the following English essay into Chinese

Because of their properties—their inherent brittleness; inability to yield, change shape, or plastically deform; sensitivity to defects; tendency to break at any point of high stress concentration; microstructures, which possess flaws that can cause failure without being detected beforehand—ceramics require a design model that calls for greater degrees of accuracy in the design and manufacture of ceramic parts.

扫一扫,查看更多资料

Chapter 9 Physical Properties of Materials

9.1 Introduction

Stone Age-Bronze Age-Iron Age-what's next? Some individuals have called the present era the space age or the atomic age. However, space exploration and nuclear reactors, to mention only two major examples, have only little impact on our everyday lives. Instead, electrical and electronic devices (such as radio, television, telephone, refrigerator, computers, electric light, CD players, electromotors, *etc*.) permeate our daily life to a large extent. Life without electronics would be nearly unthinkable in many parts of the world. The present era could, therefore, be called the age of electricity. However, electricity needs a medium in which to manifest itself and to be placed in service. For this reason, and because previous eras have been named after the material that had the largest impact on the lives of mankind, the present time may best be characterized by the name.

We are almost constantly in contact with electronic materials, such as conductors, insulators, semiconductors, (ferro) magnetic materials, optically transparent matter and opaque substances. The useful properties of these materials are governed and are characterized by electrons. In fact, the terms electronic materials and electronic properties should be understood in the widest possible sense, meaning to include all phenomena in which electrons participate in an active (dynamic) role. This is certainly the case for electrical, thermal, magnetic and even many optical phenomena.

In contrast to this, mechanical properties can be mainly interpreted by taking the interactions of atoms into account.

9.2 Electrical Properties of Materials

The prime objective of this section is to explore the electrical properties of materials, that is, their responses to an applied electric field. We begin with the phenomenon of electrical conduction: the parameters by which it is expressed, the mechanism of conduction by electrons, and how the electron energy band structure of a material influences its ability to conduct. These principles are extended to metals, semiconductors and insulators. Particular attention is given to the characteristics of semiconductors. Also treated are the dielectric characteristics of insulating materials. The final sections are devoted to the peculiar phenomena of ferroelectricity and piezoelectricity.

One of the principal characteristics of materials is their ability (or lack of ability) to conduct electrical current. Indeed, materials are classified by this property, that is, they are divided into conductors, semiconductors and nonconductors. (The latter are often called insulators or dielectrics). The conductivity, σ, of different materials at room temperature spans more than 25 orders of magnitude, as depicted in Figure 9.1. Moreover, if one takes the conductivity of superconductors, measured at low temperatures, into consideration, this span extends to 40 orders of magnitude (using an estimated conductivity for superconductors of about 10^{20} 1/($\Omega \cdot$ cm)). This is the largest known variation in a physical property and is only comparable to the ratio between the diameter of the universe (about 10^{26} m) and the radius of an electron (10^{-14} m).

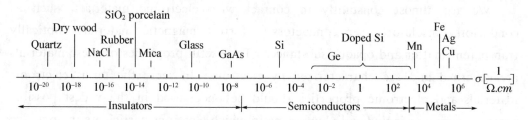

Figure 9.1 Room-temperature conductivity of various materials

(Superconductors, having conductivities of many orders of magnitude larger than copper, near 0 K, are not shown. The conductivity of semiconductors varies substantially with temperature and purity)

9.2.1 Metals and Alloys

We know that the electrons of isolated atoms (for example in a gas) can be considered to orbit at various distances about their nuclei. These orbits constitute different energies. Specifically, the larger the radius of an orbit, the larger the excitation energy of the electron. This fact is often represented in a somewhat different fashion by stating that the electrons are distributed on different energy levels, as schematically shown on the right side of Figure 9.2. Now, these distinct energy levels, which are characteristic for isolated atoms, widen into energy bands when atoms approach each other and eventually form a solid as depicted on the left side of Figure 9.2.

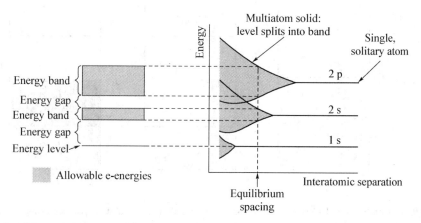

Figure 9.2 Schematic representation of energy levels (as for isolated atoms) and widening of these levels into energy bands with decreasing distance between atoms. Energy bands for a specific case are shown at the left of the diagram

The number of states within each band will equal the total of all states contributed by the N atoms. For example, an s band will consist of N states, and a p band of $3N$ states. With regard to occupancy, each energy state may accommodate two electrons, which must have oppositely directed spins. Furthermore, bands will contain the electrons that resided in the corresponding levels of the isolated atoms; for example, a $4s$ energy band in the solid will contain those isolated atom's $4s$ electrons. Of course, there will be empty bands and, possibly, bands that are only partially filled.

The electrical properties of a solid material are a consequence of its electron band structure, that is, the arrangement of the outermost electron bands and the way in which they are filled with electrons. Notably energy bands and their occupation by electrons were introduced. It was suggested that metals were distinct from insulators, and that semiconductors occupied a middle ground between the two. Simple band diagram representations in Figure 9.3 graphically distinguish these three types of solids. Semiconductors have a relatively small energy gap, E_g, of ~1 eV separating the nearly full valence and almost empty conduction bands. Insulators have a larger energy gap (5 ~ 10 eV) between the even fuller valence band and virtually unoccupied conduction band. The size of the energy gap may be viewed as a barrier to electrical conduction and that is why insulators are poor conductors. Conversely, metals have no effective energy gap because valence and conduction bands either overlap (Figure 9.3(a)) or the conduction band is only partially filled. Therefore, they are good conductors.

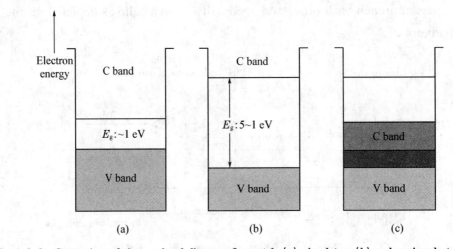

Figure 9.3 Comparison of electron band diagrams for metals (a), insulators (b) and semiconductors (c)

We may now return to the conductivity. In short, according to quantum theory, only those materials that possess partially filled electron bands are capable of conducting an electric current. Electrons can then be lifted slightly above the Fermi energy[①] into an allowed and unfilled energy state. This permits them to be accelerated by an electric field, thus producing a current. Second, only those electrons that are close to the Fermi energy participate in the electric conduction. (The classical electron theory taught us instead that all free electrons would contribute to the current). Third, the number of electrons near the Fermi energy depends on the density of available electron states.

9.2.2 Semiconductors

Semiconductors such as silicon or germanium are neither good conductors nor good insulators as seen in Figure 9.1. This may seem to make semiconductors to be of little interest. Their usefulness results, however, from a completely different property, namely, that extremely small amounts of certain impurity elements, which are called dopants, remarkably change the electrical behavior of semiconductors. Indeed, semiconductors have been proven in recent years to be the lifeblood of a multibillion dollar industry which prospers essentially from this very feature. Silicon, the major species of semiconducting materials, is today the single most researched element. Silicon is abundant (28 % of the earth's crust consists of it); the raw material (SiO_2 or sand) is inexpensive; Si forms a natural, insulating oxide; its heat conduction is reasonable; it is nontoxic; and it is stable against environmental influences.

9.2.2.1 Intrinsic Semiconductors

The properties of semiconductors are commonly explained by making use of the already introduced electron band structure which is the result of quantum-mechanical considerations. In simple terms, the electrons are depicted to reside in certain allowed energy regions. Specifically, Figure 9.3(c) and Figure 9.4 depict two electron bands, the lower of which, at 0 K, is completely filled with valence electrons. This band is appropriately called the valence band. It is separated by a small gap (about 1.1 eV for Si) from the conduction band, which, at 0 K, contains no electrons. Further, quantum mechanics stipulates that electrons essentially are not allowed to reside in the gap between these bands (called the forbidden band). Since the filled valence band possesses no allowed empty energy states in which the electrons can be thermally excited (and then accelerated in an electric field), and since the conduction band contains no electrons at all, silicon, at 0 K, is an insulator.

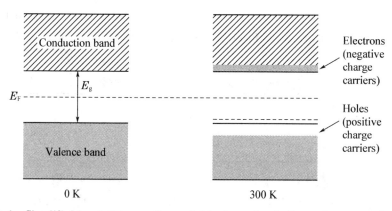

Figure 9.4 Simplified band diagrams for an intrinsic semiconductor such as pure silicon at two different temperatures. The dark shading symbolizes electrons

9.2.2.2 Extrinsic Semiconductors

The number of electrons in the conduction band can be considerably increased by adding, for example, to silicon small amounts of elements from Group V of the Periodic Table called donor atoms. Dopants such as phosphorous or arsenic are commonly utilized, which are added in amounts of, for example, 0.0001 %. These dopants replace some regular lattice atoms in a substitutional manner. Since phosphorous has five valence electrons, that is, one more than silicon, the extra electron, called the donor electron, is only loosely bound. The binding energy of phosphorous donor electrons in a silicon matrix, for example, is about 0.045 eV. Thus, the donor electrons can be disassociated from their nuclei by only a slight increase in thermal energy. Indeed, at room temperature all donor electrons have already been excited into the conduction band.

It is common to describe this situation by introducing into the forbidden band so-called donor levels, which accommodate the donor electrons at 0 K; see Figure 9.5(a). The distance between the donor level and the conduction band represents the energy that is needed to transfer the extra electrons into the conduction band (e.g., 0.045 eV for P in Si). The electrons that have been excited from the donor levels into the conduction band are free and can be accelerated in an electric field as shown in Figure 9.2 and Figure 9.4. Since the conduction mechanism in semiconductors with donor impurities is predominated by negative charge carriers, one calls these materials n-type semiconductors. Similar considerations may be carried out with respect to impurities from the third group of the Periodic Table (B, Al, Ga and In). They are deficient in one electron compared to silicon and therefore tend to accept an electron. The conduction mechanism in these semiconductors with acceptor impurities is thus predominated by positive charge carriers (holes) which are introduced from the acceptor levels (Figure 9.5(b)) into

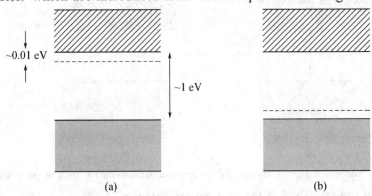

Figure 9.5 Donor (a) and acceptor (b) levels in extrinsic semiconductors

Chapter 9 Physical Properties of Materials

the valence band. They are therefore called *p*-type semiconductors. In other words, the conduction in *p*-type semiconductors under the influence of an external electric field occurs in the valence band and is predominated by holes.

9.2.2.3 Compound Semiconductors

Compounds made of Group Ⅲ and Group Ⅴ elements, such as gallium arsenide, have similar semiconducting properties as the Group Ⅳ materials silicon or germanium. GaAs is of some technical interest because of its above-mentioned wider band gap, and because of its larger electron mobility, which aids in high speed applications. Further, the ionization energies of donor and acceptor impurities in GaAs are one order of magnitude smaller than in silicon, which ensures complete electron (and hole) transfer from the donor (acceptor) levels into the conduction (valence) bands, even at relatively low temperatures. However, GaAs is about ten times more expensive than Si and its heat conduction is smaller. Other compound semiconductors include Ⅱ-Ⅵ combinations such as ZnO, ZnS, ZnSe or CdTe and Ⅳ-Ⅵ materials such as PbS, PbSe or PbTe. Silicon carbide, a Ⅳ-Ⅳ compound, has a band gap of 3 eV and can thus be used up to 700 ℃ before intrinsic effects set in. The most important application of compound semiconductors is, however, for opto-electronic[2] purposes (*e.g.* for light-emitting[3] diodes and lasers).

9.2.3 Ionic Ceramics and Polymers

Most polymers and ionic ceramics are insulating materials at room temperature and, therefore, have electron energy band structures similar to that represented in Figure 9.3(c); a filled valence band is separated from an empty conduction band by a relatively large band gap, usually greater than 2 eV. Thus, at normal temperatures only very few electrons may be excited across the band gap by the available thermal energy, which accounts for the very small values of conductivity; Table 9.1 gives the room temperature electrical conductivities of several of these materials. Of course, many materials are utilized on the basis of their ability to insulate, and thus a high electrical resistivity is desirable. With rising temperature, insulating materials experience an increase in electrical conductivity, which may ultimately be greater than that for semiconductors.

Table 9.1 Typical room-temperature electrical condtictivities for 13 nonmetallic materials

Material	Electrical conductivity / $(\Omega \cdot m^{-1})$
Graphite	$3 \times 10^4 \sim 2 \times 10^5$
Ceramics	—
Concrets (dry)	10^{-9}
Soda-lime glass	$10^{-10} \sim 10^{-11}$
Porcelain	$10^{-10} \sim 10^{-12}$
Borosilicate glass	$\sim 10^{-13}$
Aluminum oxide	$< 10^{-13}$
Fused silica	$< 10^{-18}$
Polymers	—
Phenol-formaldehyde	$10^{-9} \sim 10^{-10}$
Poly (methyl methacrylate)	$< 10^{-12}$
Nylon 6, 6	$10^{-12} \sim 10^{-13}$
Polystyrene	$< 10^{-14}$
Polyethylene	$10^{-15} \sim 10^{-17}$
Polytetrafluoroethylene	$< 10^{-17}$

9.3 Thermal Properties of Materials

By "thermal property" is meant the response of a material to the application of heat. As a solid absorbs energy in the form of heat, its temperature rises and its dimensions increase. The energy may be transported to cooler regions of the specimen if temperature gradients exist, and ultimately, the specimen may melt. Heat capacity, thermal expansion and thermal conductivity are properties that are often critical in the practical utilization of solids.

9.3.1 Heat Capacity

A solid material, when heated, experiences an increase in temperature signifying that some energy has been absorbed. Heat capacity is a property that is indicative of a material's ability to absorb heat from the external surroundings; it represents the amount of energy required to produce a unit temperature rise. In mathematical terms, the heat capacity C is expressed as follows:

$$C = \frac{dQ}{dT} \qquad (9-1)$$

where dQ is the energy required to produce a dT temperature change. Ordinarily, heat capacity is specified per mole of material (e. g., J/mol · K, or cal/mol · K). Specific heat (often denoted by a lowercase c) is sometimes used; this represents the heat capacity per unit mass and has various units (J/kg · K, cal/g · K, Btu/lbm · °F).

There are really two ways in which this property may be measured, according to the environmental conditions accompanying the transfer of heat. One is the heat capacity while maintaining the specimen volume constant, C_v; the other is for constant external pressure, which is denoted C_p. The magnitude of C_p is almost always greater than C_v; however, this difference is very slight for most solid materials at room temperature and below.

9.3.2 Thermal Expansion

Most solid materials expand upon heating and contract when cooled. The change in length with temperature for a solid material may be expressed as follows:

$$\frac{l_f - l_0}{l_0} = \alpha_1 (T_f - T_0) \tag{9-2a}$$

or

$$\frac{\Delta l_0}{l_0} = \alpha_1 \Delta T \tag{9-2b}$$

where l_0 and l_f represent, respectively, initial and final lengths with the temperature change from to the T_0 to T_f, he parameter α_1 is called the linear coefficient of thermal expansion; it is a material property that is indicative of the extent to which a material expands upon heating, and has units of reciprocal temperature (°C^{-1} or °F^{-1}). Of course, heating or cooling affects all the dimensions of a body, with a resultant change in volume. Volume changes with temperature may be computed from

$$\frac{\Delta V}{V} = \alpha_v \Delta T \tag{9-3}$$

where ΔV and V_0 are the volume change and the original volume, respectively, and α_v symbolizes the volume coefficient of thermal expansion. In many materials, the value of α_v is anisotropic; that is, it depends on the crystallographic direction along which it is measured. For materials in which the thermal expansion is isotropic, α_v is approximately $3\alpha_1$.

9.3.2.1 Metals

Linear coefficients of thermal expansion for some of the common metals range between about 5×10^{-6} and 25×10^{-6} $°C^{-1}$; these values are intermediate in magnitude between those for ceramic and polymeric materials. As the following materials of importance piece explains, several low-expansion and controlled-expansion metal alloys have been developed, which are used in applications requiring dimensional stability with temperature variations.

9.3.2.2 Ceramics

Relatively strong interatomic bonding forces are found in many ceramic materials as reflected in comparatively low coefficients of thermal expansion; values typically range between about 0.5×10^{-6} and 15×10^{-6} $°C^{-1}$. For noncrystalline ceramics and also those having cubic crystal structures, α_l is isotropic. Otherwise, it is anisotropic; and, in fact, some ceramic materials, upon heating, contract in some crystallographic directions while expanding in others. For inorganic glasses, the coefficient of expansion is dependent on composition. Fused silica (high-purity SiO_2 glass) has a small expansion coefficient, 0.4×10^{-6} $°C^{-1}$. This is explained by a low atomic packing density such that interatomic expansion produces relatively small macroscopic dimensional changes.

Ceramic materials that are to be subjected to temperature changes must have coefficients of thermal expansion that are relatively low, and in addition, isotropic. Otherwise, these brittle materials may experience fracture as a consequence of nonuniform dimensional changes in what is termed thermal shock.

9.3.2.3 Polymers

Some polymeric materials experience very large thermal expansions upon heating as indicated by coefficients that range from approximately 50×10^{-6} to 400×10^{-6} $°C^{-1}$. The highest α_l values are found in linear and branched polymers because the secondary intermolecular bonds are weak, and there is a minimum of cross-linking. With increased crosslinking, the magnitude of the expansion coefficient diminishes; the lowest coefficients are found in the thermosetting network polymers such as phenol-formaldehyde, in which the bonding is almost entirely covalent.

9.3.3 Thermal Conductivity

Thermal conduction is the phenomenon by which heat is transported from high- to low-temperature regions of a substance. The property that characterizes the ability of a material to transfer heat is the thermal conductivity.

Chapter 9 Physical Properties of Materials

Heat is transported in solid materials by both lattice vibration waves (phonons) and free electrons. Free or conducting electrons participate in electronic thermal conduction. To the free electrons in a hot region of the specimen is imparted a gain in kinetic energy. They then migrate to colder areas, where some of this kinetic energy is transferred to the atoms themselves (as vibrational energy) as a consequence of collisions with phonons or other imperfections in the crystal. The relative contribution of to the total thermal conductivity increases with increasing free electron concentrations, since more electrons are available to participate in this heat transference process.

9.3.4 Thermal Stresses

Thermal stresses are stresses induced in a body as a result of changes in temperature. An understanding of the origins and nature of thermal stresses is important because these stresses can lead to fracture or undesirable plastic deformation.

9.3.4.1 Stresses Resulting from Restrained Thermal Expansion and Contraction

Let us first consider a homogeneous and isotropic solid rod that is heated or cooled uniformly; that is, no temperature gradients are imposed. For free expansion or contraction, the rod will be stress free. If, however, axial motion of the rod is restrained by rigid end supports, thermal stresses will be introduced. The magnitude of the stress σ resulting from a temperature change from T_0 to T_f is

$$\sigma = E\alpha_1(T_0 - T_f) = E\alpha_1 \Delta T \tag{9-4}$$

where E is the modulus of elasticity and α_1 is the linear coefficient of thermal expansion. Upon heating ($T_f > T_0$), the stress is compressive ($\sigma < 0$) since rod expansion has been constrained. Of course, if the rod specimen is cooled ($T_f < T_0$), a tensile stress will be imposed ($\sigma > 0$). Also, the stress is the same as the stress that would be required to elastically compress (or elongate) the rod specimen back to its original length after it had been allowed to freely expand (or contract) with the $T_0 - T_f$ temperature change.

9.3.4.2 Stresses Resulting from Temperature Gradients

When a solid body is heated or cooled, the internal temperature distribution will depend on its size and shape, the thermal conductivity of the material and the rate of temperature change. Thermal stresses may be established as a result of temperature gradients across a body, which are frequently caused by rapid heating or cooling, in that the outside changes temperature more rapidly than the interior; differential dimensional changes serve to restrain the free expansion or contraction of adjacent volume elements within the piece. For example, upon heating, the exterior of a specimen is hotter and, therefore, will have expanded more than the interior regions. Hence, compressive surface stresses are induced and are balanced by tensile interior stresses. The interior-exterior stress conditions are reversed for rapid cooling such that the surface is put into a state of tension.

9.3.4.3 Thermal Shock of Brittle Materials

For ductile metals and polymers, alleviation of thermally induced stresses may be accomplished by plastic deformation. However, the nonductility of most ceramics enhances the possibility of brittle fracture from these stresses. Rapid cooling of a brittle body is more likely to inflict such thermal shock than heating, since the induced surface stresses are tensile. Crack formation and propagation from surface flaws are more probable when an imposed stress is tensile.

The capacity of a material to withstand this kind of failure is termed its thermal shock resistance. For a ceramic body that is rapidly cooled, the resistance to thermal shock depends not only on the magnitude of the temperature change, but also on the mechanical and thermal properties of the material. The thermal shock resistance is best for ceramics that have high fracture strengths σ_f and high thermal conductivities, as well as low moduli of elasticity and low coefficients of thermal expansion.

The resistance of many materials to this type of failure may be approximated by a thermal shock resistance parameter TSR:

$$TSR \cong \frac{\sigma_1 k}{E_{\alpha 1}} \tag{9-5}$$

Thermal shock may be prevented by altering the external conditions to the degree that cooling or heating rates are reduced and temperature gradients across a body are minimized. Modification of the thermal and/or mechanical characteristics in Equation 9-5 may also enhance the thermal shock resistance of a material. Of these parameters, the coefficient of thermal expansion is probably most easily changed and controlled. For example, common soda-lime glasses, which have an α_1 of approximately $9 \times 10^{-6} \ ℃^{-1}$, are particularly susceptible to thermal shock, as anyone who has baked can probably attest. Reducing the CaO and Na_2O contents while at the same time adding B_2O_3 in sufficient quantities to form borosilicate (or pyrex) glass will reduce the coefficient of expansion to about $3 \times 10^{-6} \ ℃^{-1}$; this material is entirely suitable for kitchen oven heating and cooling cycles. The introduction of some relatively large pores or a ductile second phase may also improve the thermal shock characteristics of a material; both serve to impede the propagation of thermally induced cracks.

It is often necessary to remove thermal stresses in ceramic materials as a means of improving their mechanical strengths and optical characteristics. This may be accomplished by an annealing heat treatment.

9.4 Magnetic Properties of Materials

Magnetism, the phenomenon by which materials assert an attractive or repulsive force or influence on other materials, has been known for thousands of years. However, the underlying principles and mechanisms that explain the magnetic phenomenon are complex and subtle, and their understanding has eluded scientists until relatively recent times. Many of our modern technological devices rely on magnetism and magnetic materials; these include electrical power generators and transformers, electric motors, radio, television, telephones, computers, and components of sound and video reproduction systems.

Iron, some steels, and the naturally occurring mineral lodestone are well-known examples of materials that exhibit magnetic properties. Not so familiar, however, is the fact that all substances are influenced to one degree or another by the presence of a magnetic field. This chapter provides a brief description of the origin of magnetic fields and discusses the various magnetic field vectors and magnetic parameters; the phenomena of diamagnetism, paramagnetism, ferromagnetism and ferrimagnetism; some of the different magnetic materials; and the phenomenon of superconductivity.

9.4.1 Diamagnetism, Paramagnetism and Ferromagnetism

Diamagnetism is a very weak form of magnetism that is nonpermanent and persists only while an external field is being applied. It is induced by a change in the orbital motion of electrons due to an applied magnetic field. The magnitude of the induced magnetic moment is extremely small, and in a direction opposite to that of the applied field. Thus, the relative permeability μ_r is less than unity (however, only very slightly), and the magnetic susceptibility χ_m is negative; that is, the magnitude of the B field within a diamagnetic solid is less than that in a vacuum. The volume susceptibility for diamagnetic solid materials is on the order of -10^{-5}.

When placed between the poles of a strong electromagnet, diamagnetic materials are attracted toward regions where the field is weak.

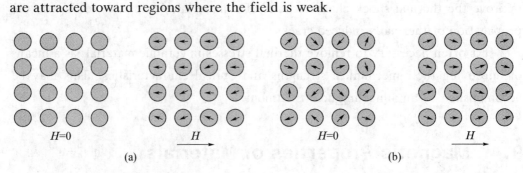

Figure 9.6 The atomic dipole configuration for a diamagnetic material with and without a magnetic field (a). In the absence of an external field, no dipoles exist; in the presence of a field, dipoles are induced that are aligned opposite to the field direction. Atomic dipole configuration with and without an external magnetic field for a paramagnetic material (b)

Figure 9.6 (a) illustrates schematically the atomic magnetic dipole configurations for a diamagnetic material with and without an external field; here, the arrows represent atomic dipole moments, whereas for the preceding discussion, arrows denoted only electron moments. The dependence of B on the external field H for a material that exhibits diamagnetic behavior is presented in Figure 9.7. Table 9.2 gives the susceptibilities of several diamagnetic materials. Diamagnetism is found in all materials; but because it is so weak, it can be observed only when other types of magnetism are totally absent. This form of magnetism is of no practical importance.

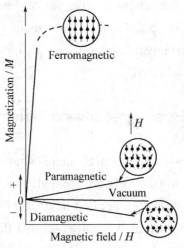

Figure 9.7 Schematic representation of the magnetization M versus the magnetic field strength H for ferromagnetic, paramagnetic and diamagnetic materials. The magnetic susceptibility is positive in paramagnets and ferromagnets, and negative in diamagnets

Chapter 9 Physical Properties of Materials

For some solid materials, each atom possesses a permanent dipole moment by virtue of incomplete cancellation of electron spin and/or orbital magnetic moments. In the absence of an external magnetic field, the orientations of these atomic magnetic moments are random, such that a piece of material possesses no net macroscopic magnetization. These atomic dipoles are free to rotate, and paramagnetism results when they preferentially align, by rotation, with an external field as shown in Figure 9.6(b). These magnetic dipoles are acted on individually with no mutual interaction between adjacent dipoles. Inasmuch as the dipoles align with the external field, they enhance it, giving rise to a relative permeability μ_r that is greater than unity, and to a relatively small but positive magnetic susceptibility. Susceptibilities for paramagnetic materials range from about 10^{-5} to 10^{-2} (Table 9.2). A schematic B-versus-H curve for a paramagnetic material is also shown in Figure 9.7.

Table 9.2 Room-temperature magnetic susceptibilities for diamagnetic and paramagnetic materials

Diamagnetics		Paramagnetics	
Materials	Susceptibility χ_m (volume) (SI units)	Materials	Susceptibility χ_m (volume) (SI units)
Aluminum oxide	-1.81×10^{-5}	Aluminum	2.07×10^{-5}
Copper	-0.96×10^{-5}	Chromium	3.13×10^{-4}
Gold	-3.44×10^{-5}	Chromium chloride	1.51×10^{-3}
Mercury	-2.85×10^{-5}	Manganese sulfate	3.70×10^{-3}
Silicon	-0.41×10^{-5}	Molybdenum	1.19×10^{-4}
Silver	-2.38×10^{-5}	Sodium	8.48×10^{-6}
Sodium chloride	-1.41×10^{-5}	Titanium	1.81×10^{-4}
Zinc	-1.56×10^{-5}	Zirconium	1.09×10^{-4}

Both diamagnetic and paramagnetic materials are considered to be nonmagnetic because they exhibit magnetization only when in the presence of an external field. Also, for both, the flux density B within them is almost the same as it would be in a vacuum.

Certain metallic materials possess a permanent magnetic moment in the absence of an external field, and manifest very large and permanent magnetizations. These are the characteristics of ferromagnetism, and they are displayed by the transition metals iron (as BCC ferrite), cobalt, nickel, and some of the rare earth metals such as gadolinium (Gd). Magnetic susceptibilities as high as 10^6 are possible for ferromagnetic materials. Consequently, $H \ll M$, and from Equation 9.6 we write

$$B \cong \mu_0 M \qquad (9-6)$$

Chapter 9 Physical Properties of Materials

Permanent magnetic moments in ferromagnetic materials result from atomic magnetic moments due to electron spin-uncancelled electron spins as a consequence of the electron structure. There is also an orbital magnetic moment contribution that is small in comparison to the spin moment. Furthermore, in a ferromagnetic material, coupling interactions cause net spin magnetic moments of adjacent atoms to align with one another, even in the absence of an external field. The origin of these coupling forces is not completely understood, but it is thought to arise from the electronic structure of the metal. This mutual spin alignment exists over relatively large volume regions of the crystal called domains.

The maximum possible magnetization, or saturation magnetization M_s of a ferromagnetic material represents the magnetization that results when all the magnetic dipoles in a solid piece are mutually aligned with the external field; there is also a corresponding saturation flux density B_s. The saturation magnetization is equal to the product of the net magnetic moment for each atom and the number of atoms present. For each of iron, cobalt and nickel, the net magnetic moments per atom are 2.22, 1.72 and 0.60 Bohr magnetons[④], respectively.

9.4.2 Antiferromagnetism and Ferrimagnetism

9.4.2.1 Antiferromagnetism

This phenomenon of magnetic moment coupling between adjacent atoms or ions occurs in materials other than those that are ferromagnetic. In one such group, this coupling results in an antiparallel alignment; the alignment of the spin moments of neighboring atoms or ions in exactly opposite directions is termed antiferroniagnetism. Manganese oxide (MnO) is one material that displays this behavior. Manganese oxide is a ceramic material that is ionic in character, having both Mn^{2+} and O^{2-} ions. No net magnetic moment is associated with the O^{2-} ions, since there is a total cancellation of both spin and orbital moments. However, the Mn^{2+} ions possess a net magnetic moment that is predominantly of spin origin. These Mn^{2+} ions are arrayed in the crystal structure such that the moments of adjacent ions are antiparallel. This arrangement is represented schematically in Figure 9.8. Obviously, the opposing magnetic moments cancel one another, and, as a consequence, the solid as a whole possesses no net magnetic moment.

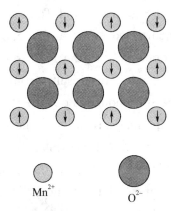

Figure 9.8 Schematic representation of antiparallel alignment of spin magnetic moments for antiferromagnetic manganese oxide

9.4.2.2 Ferrimagnetism

Some ceramics also exhibit a permanent magnetization, termed ferrimagnetism. The macroscopic magnetic characteristics of ferromagnets and ferrimagnets are similar; the distinction lies in the source of the net magnetic moments. The principles of ferrimagnetism are illustrated with the cubic ferrites. These ionic materials may be represented by the chemical formula MFe_2O_4, in which M represents any one of several metallic elements. The prototype ferrite is Fe_3O_4, the mineral magnetite, sometimes called lodestone.

The formula for Fe_3O_4 may be written as $Fe^{2+}O^{2-}(Fe^{3+})_2(O^{2-})_3$ in which the Fe ions exist in both +2 and +3 valence states in the ratio of 1:2. A net spin magnetic moment exists for each Fe^{2+} and Fe^{3+} ion, which corresponds to 4 and 5 Bohr magnetons, respectively, for the two ion types. Furthermore, the O^{2-} ions are magnetically neutral. There are antiparallel spin-coupling interactions between the Fe ions, similar in character to antiferromagnetism. However, the net ferromagnetic moment arises from the incomplete cancellation of spin moments.

Cubic ferrites have the inverse spinel crystal structure, which is cubic in symmetry, and similar to the spinel structure. The inverse spinel crystal structure might be thought of as having been generated by the stacking of closepacked planes of O^{2-} ions. For one, the coordination number is 4 (tetrahedral coordination); that is, each Fe ion is surrounded by four oxygen nearest neighbors. For the other, the coordination number is 6 (octahedral coordination). With this inverse spinel structure, half the trivalent (Fe^{3+}) ions are situated in octahedral positions, the other half, in tetrahedral positions. The divalent Fe^{2+} ions are all located in octahedral positions. The critical factor is the arrangement of the spin moments of

Chapter 9 Physical Properties of Materials

the Fe ions, as represented in Figure 9.9 and Table 9.3. The spin moments of all the Fe^{3+} ions in the octahedral positions are aligned parallel to one another; however, they are directed oppositely to the Fe^{3+} ions disposed in the tetrahedral positions, which are also aligned. This results from the antiparallel coupling of adjacent iron ions. Thus, the spin moments of all Fe^{3+} ions cancel one another and make no net contribution to the magnetization of the solid. All the Fe^{2+} ions have their moments aligned in the same direction; this total moment is responsible for the net magnetization (see Table 9.4). Thus, the saturation magnetization of a ferrimagnetic solid may be computed from the product of the net spin magnetic moment for each Fe^{2+} ion and the number of Fe^{2+} ions; this would correspond to the mutual alignment of all the Fe^{2+} ion magnetic moments in the Fe_3O_4 specimen.

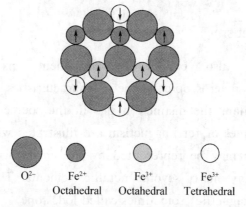

Figure 9.9 Schematic diagram showing the spin magnetic moment configuration

Table 9.3 The distribution of spin magnetic moments for Fe^{2+} and Fe^{3+} ions in a unit cell Fe_3O_4

Cation	Ocathedral lattice site	Tetrahedral lattice site	Net magnetic moment
Fe^{3+}	↑ ↑ ↑ ↑ ↑ ↑ ↑ ↑	↓ ↓ ↓ ↓ ↓ ↓ ↓ ↓	Complete cancellation
Fe^{2+}	↑ ↑ ↑ ↑ ↑ ↑ ↑ ↑	—	↑ ↑ ↑ ↑ ↑ ↑ ↑ ↑

Table 9.4 Net magnetic moments for six cations

Cation	Net spin magnetic moment (Bobr magnetons)
Fe^{3+}	5
Fe^{2+}	4
Mn^{2+}	5
Co^{2+}	3
Ni^{2+}	2
Cu^{2+}	1

Cubic ferrites having other compositions may be produced by adding metallic ions that substitute for some of the iron in the crystal structure. Again, from the ferrite chemical formula, $M^{2+}O^{2-}(Fe^{3+})_2(O^{2-})_3$, in addition to Fe^{2+}, M^{2+} may represent divalent ions such as Ni^{2+}, Mn^{2+}, Co^{2+} and Cu^{2+}, each of which possesses a net spin magnetic moment different from 4; several are listed in Table 9.4. Thus, by adjustment of composition, ferrite compounds having a range of magnetic properties may be produced. For example, nickel ferrite has the formula $NiFe_2O_4$. Other compounds may also be produced containing mixtures of two divalent metal ions such as $(Mn, Mg)Fe_2O_4$, in which the $Mn^{2+}:Mg^{2+}$ ratio may be varied; these are called mixed ferrites.

Ceramic materials other than the cubic ferrites are also ferrimagnetic; these include the hexagonal ferrites and garnets. Hexagonal ferrites have a crystal structure similar to the inverse spinel, with hexagonal symmetry rather than cubic. The chemical formula for these materials may be represented by $AB_{12}O_{19}$, in which A is a divalent metal such as barium, lead or strontium, and B is a trivalent metal such as aluminum, gallium, chromium or iron. The two most common examples of the hexagonal ferrites are $PbFe_{12}O_{19}$ and $BaFe_{12}O_{19}$.

The garnets have a very complicated crystal structure, which may be represented by the general formula $M_3Fe_5O_{12}$; here, M represents a rare earth ion such as samarium, europium, gadolinium or yttrium. Yttrium iron garnet ($Y_3Fe_5O_{12}$), sometimes denoted YIG, is the most common material of this type.

The saturation magnetizations for ferrimagnetic materials are not as high as for ferromagnets. On the other hand, ferrites, being ceramic materials, are good electrical insulators. For some magnetic applications, such as high-frequency transformers, a low electrical conductivity is most desirable.

9.4.3 The Influence of Temperature on Magnetic Behavior

Temperature can also influence the magnetic characteristics of materials. Recall that raising the temperature of solid results in an increase in the magnitude of the thermal vibrations of atoms. The atomic magnetic moments are free to rotate; hence, with rising temperature, the increased thermal motion of the atoms tends to randomize the directions of any moments that may be aligned.

For ferromagnetic, antiferromagnetic and ferrimagnetic materials, the atomic thermal motions counteract the coupling forces between the adjacent atomic dipole moments, causing some dipole misalignment, regardless of whether an external field is present. This results in a decrease in the saturation magnetization for both ferro and ferrimagnets. The saturation magnetization is a maximum at 0 K, at which temperature the thermal vibrations are a minimum. With increasing temperature, the saturation magnetization diminishes gradually and then abruptly drops to zero at what is called the Curie temperature[5] T_c.

Antiferromagnetism is also affected by temperature; this behavior vanishes at what is called the Neel temperature[6]. At temperatures above this point, antiferromagnetic materials also become paramagnetic.

9.4.4 Domains, Hysteresis and Magnetic Anisotropy

Any ferromagnetic or ferrimagnetic material that is at a temperature below is composed of small-volume regions in which there is a mutual alignment in the same direction of all magnetic dipole moments. Such a region is called a domain, and each one is magnetized to its saturation magnetization. Adjacent domains are separated by domain boundaries or walls, across which the direction of magnetization gradually changes. Normally, domains are microscopic in size, and for a polycrystalline specimen, each grain may consist of more than a single domain. Thus, in a macroscopic piece of material, there will be a large number of domains, and all may have different magnetization orientations. The magnitude of the M field for the entire solid is the vector sum of the magnetizations of all the domains, each domain contribution being weighted by its volume fraction. For an unmagnetized specimen, the appropriately weighted vector sum of the magnetizations of all the domains is zero.

Flux density B and field intensity H are not proportional for ferromagnets and ferrimagnets. If the material is initially unmagnetized, then B varies as a function of H as shown in Figure 9.10. The curve begins at the origin, and as H is increased, the B field begins to increase slowly, then more rapidly, finally leveling off and becoming independent of H. This maximum value of B is the saturation flux density B_s, and the corresponding magnetization is the saturation magnetization M_s, mentioned previously. Since the permeability μ is the slope of the B-versus-H curve, note from Figure 9.10 that the permeability changes with and is dependent on H. On occasion, the slope of the B-versus-H curve at $H = 0$ is specified as a material property, which is termed the initial permeability μ_i, as indicated in Figure 9.10.

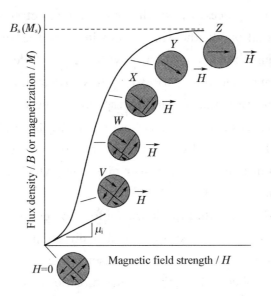

Figure 9.10 The B-versus-H behavior for a ferromagnetic or ferrimagnetic material that was initially unmagnetized. Domain configurations during several stages of magnetization are represented. Saturation flux density B_s, magnetization M_s, and initial permeability μ_i

As an H field is applied, the domains change shape and size by the movement of domain boundaries. Schematic domain structures are represented in the insets (labeled U through Z) at several points along the B-versus-H curve in Figure 9.10. Initially, the moments of the constituent domains are randomly oriented such that there is no net B (or M) field (inset U). As the external field is applied, the domains that are oriented in directions favorable to (or nearly aligned with) the applied field grow at the expense of those that are unfavorably oriented (insets V through X). This process continues with increasing field strength until the macroscopic specimen becomes a single domain, which is nearly aligned with the field (inset Y). Saturation is achieved when this domain, by means of rotation, becomes oriented with the H field (inset Z). Alteration of the domain structure with magnetic field for an iron single crystal is shown in the chapter-opening photographs for this chapter.

From saturation, point S in Figure 9.11, as the H field is reduced by reversal of field direction, the curve does not retrace its original path. A hysteresis effect is produced in which the B field lags behind the applied H field, or decreases at a lower rate. At zero H field (point R on the curve), there exists a residual B field that is called the remanence, or remanent flux density, B_r; the material remains magnetized in the absence of an external H field.

Hysteresis behavior and permanent magnetization may be explained by the motion of domain walls. Upon reversal of the field direction from saturation (point

S in Figure 9.11), the process by which the domain structure changes is reversed. First, there is a rotation of the single domain with the reversed field. Next, domains having magnetic moments aligned with the new field form and grow at the expense of the former domains. Critical to this explanation is the resistance to movement of domain walls that occurs in response to the increase of the magnetic field in the opposite direction; this accounts for the lag of B with H, or the hysteresis. When the applied field reaches zero, there is still some net volume fraction of domains oriented in the former direction, which explains the existence of the remanence B_r.

To reduce the B field within the specimen to zero (point C on Figure 9.11), an H field of magnitude-H_c must be applied in a direction opposite to that of the original field; H_c is called the coercivity, or sometimes the coercive force. Upon continuation of the applied field in this reverse direction, as indicated in the figure, saturation is ultimately achieved in the opposite sense, corresponding to point S'. A second reversal of the field to the point of the initial saturation (point S) completes the symmetrical hysteresis loop and also yields both a negative remanence ($-B_r$) and a positive coercivity ($+H_c$).

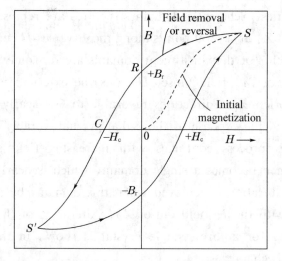

Figure 9.11 Magnetic flux density versus the magnetic field strength for a ferromagnetic material that is subjected to forward and reverse saturations (points S and S'). The hysteresis loop is represented by the solid curve; the dashed curve indicates the initial magnetization. The remanence B_r and the coercive force He are also shown

The magnetic hysteresis curves discussed in the previous section will have different shapes depending on various factors: (1) whether the specimen is a single crystal or polycrystalline; (2) if polycrystalline, any preferred orientation of the grains; (3) the presence of pores or second-phase particles; and (4) other factors

Chapter 9 Physical Properties of Materials

such as temperature and, if a mechanical stress is applied, the stress state.

9.4.5 Superconductivity

The resistivity in superconductors becomes immeasurably small or virtually zero below a critical temperature, T_c, as shown in Figure 9.12. About 27 elements, numerous alloys, ceramic materials (containing copper oxide) and organic compounds (based, for example, on selenium or sulfur) have been found to possess this property (see Table 9.5). It is estimated that the conductivity of superconductors below T_c is about 10^{20} $1/\Omega \cdot cm$.

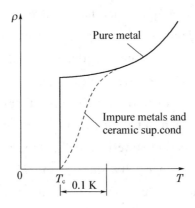

Figure 9.12 Schematic representation of the resistivity of pure and compound superconducting materials. T_c is the critical or transition temperature, below which superconductivity commences

Table 9.5 Critical temperatures of some superconducting materials

Materials	T_c/ K	Remarks
Tungsten	0.01	——
Mercury	4.15	H. K. Onnes (1911)
Sulfur-based organic superconductor	8	S. S. P. Parkin, et al. (1983)
Nb_3Sn and Nb-Ti	9	Bell Labs (1961), Type II
V_3Si	17.1	J. K. Hulm (1953)
Nb_3Ge	23.2	(1973)
La-Ba-Cu-O	40	Bednorz and müller (1986)
$YBa_2Cu_3O_7$	92	Wu, Chu, and others (1987)
$RBa_2Cu_3O_{7-x}$	~92	R= Gd, Dy, Ho, Er, Tm, Yb, Lu
$Bi_2Sr_2Ca_2Cu_3O_{10+\delta}$	113	Maeda, et al. (1988)
$Tl_2CaBa_2Cu_2O_{10+\delta}$	125	Hermann, et al. (1988)
$HgBa_2Ca_2Cu_3O_{8+\delta}$	134	R. Ott, et al. (1995)

Chapter 9 Physical Properties of Materials

The transition temperatures where superconductivity starts range from 0.01 K (for tungsten) up to about 125 K (for ceramic superconductors). Of particular interest are materials whose T_c is above 77 K, that is, the boiling point of liquid nitrogen, which is more readily available than other coolants. Among the so-called high-T_c superconductors are the 1-2-3 compounds such as $YBa_2Cu_3O_{7-x}$ whose molar ratios of rare earth to alkaline earth to copper relate as 1 : 2 : 3. Their transition temperatures range from 40 to 134 K. Ceramic superconductors have an orthorhombic, layered, perovskite crystal structure which contains two-dimensional sheets and periodic oxygen vacancies (The superconductivity exists only parallel to these layers, that is, it is anisotropic). The first superconducting material was found by H. K. Onnes in 1911 in mercury which has a T_c of 4.15 K.

A high magnetic field or a high current density may eliminate superconductivity. In type I superconductors, the annihilation of the superconducting state by a magnetic field, that is, the transition between superconducting and normal states, occurs sharply; Figure 9.13(a). The critical field strength H_c, above which superconductivity ceases, is relatively low. The destruction of the superconducting state in type II superconductors occurs instead, more gradually, i.e., in a range between H_{c1} and H_{c2}, where H_{c2} is often 100 times larger than H_{c1} (Figure 9.13(b)). In the interval between H_{c1} and H_{c2}, normal conducting areas, called vortices, and superconducting regions are interspersed. The terms "type I" and "type II" superconductors are occasionally also used when a distinction between abrupt and gradual transition with respect to temperature is described; see Figure 9.13. In alloys and ceramic superconductors, a temperature spread of about 0.1 K has been found whereas pure gallium drops its resistance within 10^{-5} K.

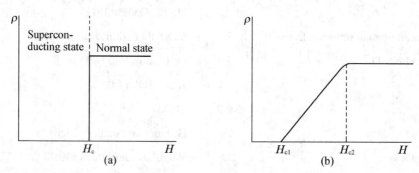

Figure 9.13 Schematic representation of the resistivity of (a) type I (or soft) and (b) type II (or hard) superconductors in an external magnetic field. The solids behave like normal conductors above H_c and H_{c2}, respectively

Type II superconductors are utilized for strong electromagnets employed, for example, in magnetic resonance imaging devices (used in medicine), high-energy particle accelerators, and electric power storage devices. Further potential applications are lossless power transmission lines; high-speed levitation trains; faster, more compact computers; and switching devices, called cryotrons, which are based on the destruction of the superconducting state in a strong magnetic field. Despite their considerably higher transition temperatures, ceramic superconductors have not yet revolutionized current technologies, mainly because of their still relatively low T_c, their brittleness, their relatively small capability to carry high current densities, and their environmental instability. These obstacles may be overcome eventually, however, by using other materials, for example, compounds based on bismuth, *etc.*, or by producing thin-film superconductors. At present, most superconducting electromagnets are manufactured by using niobium-titanium alloys which are ductile and thus can be drawn into wires.

Superconductivity is basically an electrical phenomenon; however, its discussion has been deferred to this point because there are magnetic implications relative to the superconducting state, and, in addition, superconducting materials are used primarily in magnets capable of generating high fields.

As most high-purity metals are cooled down to temperatures nearing 0 K, the electrical resistivity decreases gradually, approaching some small yet finite value that is characteristic of the particular metal. There are a few materials, however, for which the resistivity, at a very low temperature, abruptly plunges from a finite value to one that is virtually zero and remains there upon further cooling. Materials that display this latter behavior are called superconductors, and the temperature at which they attain superconductivity is called the critical temperature T_c. The resistivity-temperature behaviors for superconductive and nonsuperconductive materials are contrasted in Figure 9.14. The critical temperature varies from

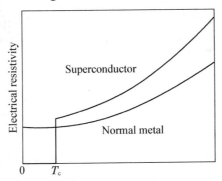

Figure 9.14 Temperature dependence of the electrical resistivity for normally conducting and superconducting materials in the vicinity of 0 K

superconductor to superconductor but lies between less than 1 K and approximately 20 K for metals and metal alloys. Recently, it has been demonstrated that some complex oxide ceramics have critical temperatures in excess of 100 K.

At temperatures below T_c, the superconducting state will cease upon application of a sufficiently large magnetic field H_c, termed the critical field which depends on temperature and decreases with increasing temperature. The same may be said for current density; that is, a critical applied current density J_c exists below which a material is superconductive.

The superconductivity phenomenon has been satisfactorily explained by means of a rather involved theory. In essence, the superconductive state results from attractive interactions between pairs of conducting electrons; the motions of these paired electrons become coordinated such that scattering by thermal vibrations and impurity atoms is highly inefficient. Thus, the resistivity, being proportional to the incidence of electron scattering, is zero.

On the basis of magnetic response, superconducting materials may be divided into two classifications designated as type I and type II. Type I materials, while in the superconducting state, are completely diamagnetic; that is, all of an applied magnetic field will be excluded from the body of material, a phenomenon known as the Meissner effect[①], which is illustrated in Figure 9.15. As H is increased, the material remains diamagnetic until the critical magnetic field H_c is reached. At this point, conduction becomes normal and complete magnetic flux penetration takes place. Several metallic elements including aluminum, lead, tin and mercury belong to the type I group.

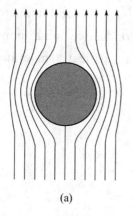 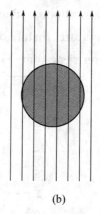

(a)   (b)

Figure 9.15 Representation of the Meissner effect. While in the superconducting state, a body of material (circle) excludes a magnetic field (arrows) from its interior (a). The magnetic field penetrates the same body of material once it becomes normally conductive (b)

Type Ⅱ superconductors are completely diamagnetic at low applied fields, and field exclusion is total. However, the transition from the superconducting state to the normal state is gradual and occurs between lower critical and upper critical fields, designated H_{c1} and H_{c2}, respectively. The magnetic flux lines begin to penetrate into the body of material at and with increasing applied magnetic field, this penetration continues; at H_{c2}, field penetration is complete. For fields between H_{c1} and H_{c2}, the material exists in what is termed a mixed state-both normal and superconducting regions are present.

Type Ⅱ superconductors are preferred over type Ⅰ for most practical applications by virtue of their higher critical temperatures and critical magnetic fields. At present, the three most commonly utilized superconductors are niobium-zirconium (Nb-Zr) and niobium-titanium (Nb-Ti) alloys and the niobium-tin intermetallic compound Nb_3Sn. Table 9.6 lists several type Ⅰ and n superconductors, their critical temperatures, and their critical magnetic flux densities.

Table 9.6 Critical temperatures and magnetic fluxes for selected superconducting materials

Material	Critical temperature (T_c) / K	Gritical magnetic flux density (B_c) / Tesla
Tungsten	0.02	0.000 1
Titanium	0.40	0.005 6
Aluminum	1.18	0.010 5
Tin	3.72	0.030 5
Mercury (α)	4.15	0.041 1
Lead	7.19	0.080 3
Compounds and alloys	——	——
Nb-Ti alloy	10.2	12
Nb-Zr alloy	10.8	11
$PbMo_6S_8$	14.0	45
V_3Ga	16.5	22
Nb_3Sn	18.3	22
Nb_3Al	18.9	32
Nb_3Ge	23.0	30
Ceramic compounds	——	——
$YBa_2Cu_3O_7$	92	——
$Bi_2Sr_2Ca_2Cu_3O_{10}$	110	——
$Tl_2Ba_2Ca_2Cu_3O_{10}$	125	——
$HgBa_2Ca_2Cu_2O_8$	153	——

Recently, the families of ceramic materials that are normally electrically insulative have been found to be superconductors with inordinately high critical temperatures. Initial research has centered on yttrium barium copper oxide, $YBa_2Cu_3O_7$, which has a critical temperature of about 92 K. This material has a complex perovskite type crystal structure. New superconducting ceramic materials reported to have even higher critical temperatures have been and are currently being developed. Several of these materials and their critical temperatures are listed in Table 9.6. The technological potential of these materials is extremely promising inasmuch as their critical temperatures are above 77 K, which permits the use of liquid nitrogen, a very inexpensive coolant in comparison to liquid hydrogen and liquid helium. These new ceramic superconductors are not without drawbacks, chief of which is their brittle nature. This characteristic limits the ability of these materials to be fabricated into useful forms such as wires.

Some of the areas being explored include: (1) electrical power transmission through superconducting materials-power losses would be extremely low, and the equipment would operate at low voltage levels; (2) magnets for high-energy particle accelerators; (3) higher-speed switching and signal transmission for computers; and (4) high-speed magnetically levitated trains, wherein the levitation results from magnetic field repulsion. The chief deterrent to the widespread application of these superconducting materials is, of course, the difficulty in attaining and maintaining extremely low temperatures. Hopefully, this problem will be overcome with the development of the new generation of superconductors with reasonably high critical temperatures.

9.5 Optical Properties of Materials

By "optical property" is meant a material's response to exposure to electromagnetic radiation and, in particular, to visible light. This chapter first discusses some of the basic principles and concepts relating to the nature of electromagnetic radiation and its possible interactions with solid materials. Next to be explored are the optical behaviors of metallic and nonmetallic materials in terms of their absorption, reflection and transmission characteristics. The final sections outline luminescence, photoconductivity and light amplification by stimulated emission of radiation (laser), the practical utilization of these phenomena and optical fibers in communications.

9.5.1 Interaction of Light with Matter

The most apparent properties of metals, their luster and their color, have been known to mankind since materials were known. Because of these properties, metals were already used in antiquity for mirrors and jewelry. The color was utilized 4 000 years ago by the ancient Chinese as a guide to determine the composition of the melt of copper alloys: the hue of a preliminary cast indicated whether the melt, from which bells or mirrors were to be made, already had the right tin content.

The German poet Goethe was probably the first one who explicitly spelled out 200 years ago in his Treatise on Color that color is not an absolute property of matter (such as the resistivity), but requires a living being for its perception and description. Goethe realized that the perceived color of a region in the visual field depends not only on the properties of light coming from that region, but also on the light coming from the rest of the visual field. Applying Goethe's findings, it was possible to explain qualitatively the color of, say, gold in simple terms. Goethe wrote: "If the color blue is removed from the spectrum, then blue, violet and green are missing and red and yellow remain". Thin gold films are bluish-green when viewed in transmission. These colors are missing in reflection. Consequently, gold appears reddish-yellow. On the other hand, Newton stated quite correctly in his "Opticks" that light rays are not colored. The nature of color remained, however, unclear.

This chapter treats the optical properties from a completely different point of view. Measurable quantities such as the index of refraction or the reflectivity and their spectral variations are used to characterize materials. In doing so, the term "color" will almost completely disappear from our vocabulary. Instead, it will be postulated that the interactions of light with the electrons of a material are responsible for the optical properties.

At the beginning of the 20th century, the study of the interactions of light with matter (black-body radiation, *etc.*) laid the foundations for quantum theory. Today, optical methods are among the most important tools for elucidating the electron structure of matter. Most recently, a number of optical devices such as lasers, photodetectors, waveguides, *etc.*, have gained considerable technological importance. They are used in telecommunication, fiber optics, CD players, laser printers, medical diagnostics, night viewing, solar applications, optical computing, and for optoelectronic purposes. Traditional utilizations of optical materials for windows, antireflection coatings, lenses, mirrors, *etc.*, should be likewise mentioned.

We perceive light intuitively as a wave (specifically, an electromagnetic wave) that travels in undulations from a given source to a point of observation. The color of the light is related to its wavelength. Many crucial experiments, such as diffraction, interference and dispersion, clearly confirm the wavelike nature of light. Nevertheless, at least since the discovery of the photoelectric effect in 1887 by Hertz, and its interpretation in 1905 by Einstein, do we know that light also has a particle nature (The photoelectric effect describes the emission of electrons from a metallic surface after it has been illuminated by light of appropriately high energy, e. g., by blue light). Interestingly enough, Newton, about 300 years ago, was a strong proponent of the particle concept of light. His original ideas, however, were in need of some refinement, which was eventually provided in 1901 by quantum theory.

Light comprises only an extremely small segment of the entire electromagnetic spectrum, which ranges from radio waves via microwaves, infrared, visible, ultraviolet, X-rays to γ-rays. Many of the considerations which will be advanced in this chapter are therefore also valid for other wavelength ranges, i. e., for radio waves or X-rays.

9.5.2 Atomic and Electronic Interactions

The optical phenomena that occur within solid materials involve interactions between the electromagnetic radiation and atoms, ions, and/or electrons. Two of the most important of these interactions are electronic polarization and electron energy transitions.

9.5.2.1 Electronic Polarization

One component of an electromagnetic wave is simply a rapidly fluctuating electric field. For the visible range of frequencies, this electric field interacts with the electron cloud surrounding each atom within its path in such a way as to induce electronic polarization, or to shift the electron cloud relative to the nucleus of the atom with each change in direction of electric field component. Two consequences of this polarization are: (1) some of the radiation energy may be absorbed, and (2) light waves are retarded in velocity as they pass through the medium.

9.5.2.2 Electron Transitions

The absorption and emission of electromagnetic radiation may involve electron transitions from one energy state to another. For the sake of this discussion, consider an isolated atom, the electron energy diagram for which is represented in Figure 9.16. An electron may be excited from an occupied state at energy E_2 to a vacant and higher-lying one, denoted E_4, by the absorption of a photon of energy. The change in energy experienced by the electron, ΔE depends on the radiation frequency as follows:

$$\Delta E = h\nu \tag{9-7}$$

where, again, h is Planck's constant[①]. At this point it is important that several concepts be understood. First, since the energy states for the atom are discrete, only specific ΔE's exist between the energy levels; thus, only photons of frequencies corresponding to the possible ΔE's for the atom can be absorbed by electron transitions. Furthermore, all of a photon's energy is absorbed in each excitation event.

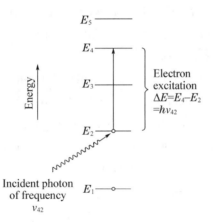

Figure 9.16 For an isolated atom, a schematic illustration of photon absorption by the excitation of an electron from one energy state to another. The energy of the photon must be exactly equal to the difference in energy between the two states ($E_4 - E_2$)

A second important concept is that a stimulated electron cannot remain in an excited state indefinitely; after a short time, it falls or decays back into its ground state, or unexcited level, with a reemission of electromagnetic radiation. Several decay paths are possible, and these are discussed later. In any case, there must be a conservation of energy for absorption and emission electron transitions.

As the ensuing discussions show, the optical characteristics of solid materials that relate to absorption and emission of electromagnetic radiation are explained in terms of the electron band structure of the material and the principles relating to electron transitions, as outlined in the preceding two paragraphs.

9.5.2.3 Optical Properties of Metals

Consider the electron energy band schemes for metals; in both cases a high-energy band is only partially filled with electrons. Metals are opaque because the incident radiation having frequencies within the visible range excites electrons into unoccupied energy states above the Fermi energy, as demonstrated in Figure 9.17(a); as a consequence, the incident radiation is absorbed, in accordance with Equation (9-6). Total absorption is within a very thin outer layer, usually less than 0.1 μm; thus only metallic films thinner than 0.1 μm are capable of transmitting visible light. All frequencies of visible light are absorbed by metals because of the continuously available empty electron states, which permit electron transitions as in Figure 9.17(a).

In fact, metals are opaque to all electromagnetic radiation on the low end of the frequency spectrum, from radio waves, through infrared, the visible, and into about the middle of the ultraviolet radiation. Metals are transparent to high-frequency (X-and-ray) radiation.

Most of the absorbed radiation is reemitted from the surface in the form of visible light of the same wavelength, which appears as reflected light; an electron transition accompanying reradiation is shown in Figure 9.17(b). The reflectivity for most metals is between 0.90 and 0.95; some small fraction of the energy from electron decay processes is dissipated as heat.

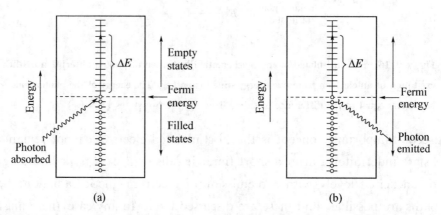

Figure 9.17 Schematic representation of the mechanism of photon absorption for metallic materials in which an electron is excited into a higher-energy unoccupied state (a). The change in energy of the electron ΔE is equal to the energy of the photon. Reemission of a photon of light by the direct transition of an electron from a high to a low energy state (b)

Since metals are opaque and highly reflective, the perceived color is determined by the wavelength distribution of the radiation that is reflected and not absorbed. A bright silvery appearance when exposed to white light indicates that the metal is highly reflective over the entire range of the visible spectrum. In other words, for the reflected beam, the composition of these reemitted photons, in terms of frequency and number, is approximately the same as for the incident beam. Aluminum and silver are two metals that exhibit this reflective behavior. Copper and gold appear red-orange and yellow, respectively, because some of the energy associated with light photons having short wavelengths is not reemitted as visible light.

9.5.2.4 Optical Properties of Nonmetals

By virtue of their electron energy band structures, nonmetallic materials may be transparent to visible light. Therefore, in addition to reflection and absorption, refraction and transmission phenomena also need to be considered.

9.5.3 Refraction, Reflection, Absorption and Transmission

Light that is transmitted into the interior of transparent materials experiences a decrease in velocity, and, as a result, is bent at the interface; this phenomenon is termed refraction.

For crystalline ceramics that have cubic crystal structures, and for glasses, the index of refraction is independent of crystallographic direction (i. e., it is isotropic). Noncubic crystals, on the other hand, have an anisotropic n; that is, the index is greatest along the directions that have the highest density of ions. Table 9.7 gives refractive indices for several glasses, transparent ceramics and polymers. Average values are provided for the crystalline ceramics in which n is anisotropic.

Table 9.7 Refractive indices for some transparent materials

Material	Average index refraction	Material	Average index refraction
Ceramics		Polymers	
Silica glass	1.458	Polytetrafluoroethylene	1.35
Borosilicate (Pyrex) glass	1.47	Polymethyl methacrylate	1.49
Soda-lime glass	1.51	Polypropylene	1.49
Quartz (SiO_2)	1.55	Polyethylene	1.51
Dense optical flint glass	1.65	Polystyrene	1.60
Spinel ($MgAl_2O_4$)	1.72		
Periclase (MgO)	1.74		
Corundum (Al_2O_3)	1.76		

When light radiation passes from one medium into another having a different index of refraction, some of the light is scattered at the interface between the two media even if both are transparent. The reflectivity R represents the fraction of the incident light that is reflected at the interface, or

$$R = \frac{I_R}{I_0} \tag{9-8}$$

where I_0 and I_R are the intensities of the incident and reflected beams, respectively.

Nonmetallic materials may be opaque or transparent to visible light; and, if transparent, they often appear colored. In principle, light radiation is absorbed in this group of materials by two basic mechanisms, which also influence the transmission characteristics of these nonmetals. One of these is electronic polarization. Absorption by electronic polarization is important only at light frequencies in the vicinity of the relaxation frequency of the constituent atoms. The other mechanism involves valence band-conduction band electron transitions, which depend on the electron energy band structure of the material.

Absorption of a photon of light may occur by the promotion or excitation of an electron from the nearly filled valence band, across the band gap, and into an empty state within the conduction band, a free electron in the conduction band and a hole in the valence band are created.

The phenomena of absorption, reflection and transmission may be applied to the passage of light through a transparent solid, as shown in Figure 9.18. For an incident beam of intensity I_0 that impinges on the front surface of a specimen of thickness l and absorption coefficient β, the transmitted intensity at the back face I_T is

$$I_T = I_0 (1 - R)^{2e^{-\beta l}} \tag{9-9}$$

where R is the reflectance; for this expression, it is assumed that the same medium exists outside both front and back faces. The derivation of Equation (9-9) is left as a homework problem.

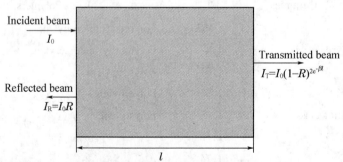

Figure 9.18 The transmission of light through a transparent medium for which there is reflection at front and back faces, as well as absorption within the medium

Thus, the fraction of incident light that is transmitted through a transparent material depends on the losses that are incurred by absorption and reflection. Again, the sum of the reflectivity R, absorptivity A, and transmissivity T, is unity according to Equation 9.5. Also, each of the variables R, A and T depends on light wavelength. For example, for light having a wavelength of 0.4 m, the fractions transmitted, absorbed, and reflected are approximately 0.90, 0.05 and 0.05, respectively. However, at 0.55 m, the respective fractions have shifted to about 0.50, 0.48 and 0.02.

9.5.4 Opacity and Translucency in Insulators

The extent of translucency and opacity for inherently transparent dielectric materials depends to a great degree on their internal reflectance and transmittance characteristics. Many dielectric materials that are intrinsically transparent may be made translucent or even opaque because of interior reflection and refraction. A transmitted light beam is deflected in direction and appears diffuse as a result of multiple scattering events. Opacity results when the scattering is so extensive that virtually none of the incident beam is transmitted, undeflected, to the back surface.

This internal scattering may result from several different sources. Polycrystalline specimens in which the index of refraction is anisotropic normally appear translucent. Both reflection and refraction occur at grain boundaries, which cause a diversion in the incident beam. This results from a slight difference in index of refraction n between adjacent grains that do not have the same crystallographic orientation.

Scattering of light also occurs in two-phase materials in which one phase is finely dispersed within the other. Again, the beam dispersion occurs across phase boundaries when there is a difference in the refractive index for the two phases; the greater this difference, the more efficient is the scattering. Glass-ceramics, which may consist of both crystalline and residual glass phases, will appear highly transparent if the sizes of the crystallites are smaller than the wavelength of visible light, and when the indices of refraction of the two phases are nearly identical (which is possible by adjustment of composition).

As a consequence of fabrication or processing, many ceramic pieces contain some residual porosity in the form of finely dispersed pores. These pores also effectively scatter light radiation.

Chapter 9 Physical Properties of Materials

9.5.5 Applications of Optical Phenomena

9.5.5.1 Luminescence

Some materials are capable of absorbing energy and then reemitting visible light in a phenomenon called luminescence. Photons of emitted light are generated from electron transitions in the solid. Energy is absorbed when an electron is promoted to an excited energy state; visible light is emitted when it falls back to a lower energy state if $1.8 \text{ eV} < h\nu < 3.1 \text{ eV}$. The absorbed energy may be supplied as higher energy electromagnetic radiation causing valence band-conduction band transitions, such as ultraviolet light, or other sources such as high energy electrons, or by heat, mechanical or chemical energy. Furthermore, luminescence is classified according to the magnitude of the delay time between absorption and reemission events. If reemission occurs for times much less than one second, the phenomenon is termed fluorescence; for longer times, it is called phosphorescence. A number of materials can be made to fluoresce or phosphoresce; these include some sulfides, oxides, tungstates and a few organic materials. Ordinarily, pure materials do not display these phenomena, and to induce them, impurities in controlled concentrations must be added.

Luminescence has a number of commercial applications. Fluorescent lamps consist of glass housing, coated on the inside with specially prepared tungstates or silicates. Ultraviolet light is generated within the tube from a mercury glow discharge, which causes the coating to fluoresce and emit white light. The picture viewed on a television screen (cathode ray tube screen) is the product of luminescence. The inside of the screen is coated with a material that fluoresces as an electron beam inside the picture tube very rapidly traverses the screen Detection of X-rays and Y-rays is also possible; certain phosphors emit visible light or glow when introduced into a beam of the radiation that is otherwise invisible.

9.5.5.2 Photoconductivity

The conductivity of semiconducting materials depends on the number of free electrons in the conduction band and also the number of holes in the valence band. Thermal energy associated with lattice vibrations can promote electron excitations in which free electrons and/or holes are created. Additional charge carriers may be generated as a consequence of photon-induced electron transitions in which light is absorbed; the attendant increase in conductivity is called photoconductivity.

Thus, when a specimen of a photoconductive material is illuminated, the conductivity increases. This phenomenon is utilized in photographic light meters. A photoinduced current is measured, and its magnitude is a direct function of the intensity of the incident light radiation, or the rate at which the photons of light strike the photoconductive material. Of course, visible light radiation must induce electronic transitions in the photoconductive material; cadmium sulfide is commonly utilized in light meters.

Sunlight may be directly converted into electrical energy in solar cells, which also employ semiconductors. The operation of these devices is, in a sense, the reverse of that for the light-emitting diode. A *p-n* junction is used in which photoexcited electrons and holes are drawn away from the junction, in opposite directions, and become part of an external current.

Optical phenomena are also applied in stimulated electron transitions such as lasers, coherent and high-intensity light beams and transmission of information such as fiber-optic technology in communications.

Notes

① Fermi energy 费米能级；费米能量
② opto-electronic 光电
③ light-emitting 发光的
④ Bohr magnetons 玻尔磁子
⑤ Curie temperature 居里温度；居里点
⑥ Neel temperature 奈耳温度
⑦ Meissner effect 迈斯纳效应
⑧ Planck's constant 普朗克氏常数

Vocabulary

absorptivity [ˌæbsɔːpˈtɪvətɪ] *n*. 吸收率；吸收能力；吸收性
align [əˈlaɪn] *vt*. 使结盟；使成一行；匹配　*vi*. 排列；排成一行
borosilicate [ˌbɔːrəʊˈsɪlɪkeɪt] *n*. 硼硅酸盐
capacity [kəˈpæsɪtɪ] *n*. 容量，能力，容积
coercivity [ˌkəʊɜːˈsɪvɪtɪ] *n*. 矫顽磁性，矫顽力
conductivity [ˌkɒndʌkˈtɪvɪtɪ] *n*. 导电性；[物][生理]传导性
conductor [kənˈdʌktə] *n*. 导体
counteract [ˌkaʊntərˈækt] *vt*. 抵消；中和；阻碍
crosslinking [kˈrɒslɪŋkɪŋ] *n*. 交叉耦合，交联
crust [krʌst] *n*. 地壳；外壳
cryotron [ˈkraɪətrɒn] *n*. [电子]冷子管；[电子]低温管
diamagnetism [ˌdaɪəˈmæɡnɪtɪzəm] *n*. 反磁性；逆磁性；反磁性学
dielectric [ˌdaɪɪˈlektrɪk] *n*. 电介质；绝缘体　*adj*. 非传导性的；诱电性的
dielectrics [ˌdaɪɪˈlektrɪks] *n*. 电介质（dielectric 的复数）
diode [ˈdaɪəʊd] *n*. [电子]二极管

Chapter 9 Physical Properties of Materials

dipole [ˈdaɪpəʊl] n. [物化]偶极；双极子

donor [ˈdəʊnə] n. 半导体

dopant [ˈdəʊpənt] n. 掺杂物；掺杂剂

electromotor [ɪˌlektrəʊˈməʊtə] n. 电动机，电气马达

europium [jʊəˈrəʊpɪəm] n. [化学]铕

excitation [ˌeksaɪˈteɪʃən] n. 激发，刺激；激励；激动

extrinsic [ɪkˈstrɪnsɪk] adj. 外在的；外来的；非固有的

ferrimagnetism [ˌferɪmægˈnetɪzəm] n. [物]铁氧体磁性

ferromagnetism [ˌferəʊˈmægnetɪzəm] n. [物]铁磁性

fluorescence [flʊəˈresəns] n. 荧光；荧光性

flux [flʌks] n. [流][机]流量；变迁；不稳定；流出 vt. 使熔融；用焊剂处理 vi. 熔化；流出

gadolinium [ˌgædəˈlɪnɪəm] n. [化学]钆

gallium [ˈgælɪəm] n. [化学]镓

garnet [ˈgɑːnɪt] n. [矿物]石榴石；深红色 adj. 深红色的；暗红色的

germanium [dʒɜːˈmeɪnɪəm] n. [化学]锗

helium [ˈhiːlɪəm] n. [化学]氦

hysteresis [ˌhɪstəˈriːsɪs] n. 迟滞现象；滞后作用，磁滞现象；滞变

insulator [ˈɪnsjʊleɪtə] n. [物]绝缘体

intrinsic [ɪnˈtrɪnsɪk] adj. 本质的，固有的

junction [ˈdʒʌŋkʃən] n. 连接，接合；交叉点；接合点

laser [ˈleɪzə] n. 激光

layered [ˈleɪəd] adj. 分层的；层状的

lodestone [ˈləʊdstəʊn] n. 天然磁石；吸引人的东西（等于 loadstone）

lowercase [ˌləʊəˈkeɪs] n. 小写字母；小写字体

luminescence [ˌluːmɪˈnesəns] n. [光]发冷光

magnetite [ˈmægnɪtaɪt] n. [矿物]磁铁矿

magneton [ˈmægnɪtɒn] n. 磁子

mobility [məʊˈbɪlətɪ] n. 移动性；机动性；[电子]迁移率

mole [məʊl] 摩尔（单位）

mutual [ˈmjuːtʃʊəl] adj. 共同的；相互的，彼此的

niobium [naɪˈəʊbɪəm] n. [化学]铌

nonconductor [ˌnɒnkənˈdʌktə] n. [物]绝缘体；[物]非导体

nontoxic [nɒnˈtɒksɪk] adj. 无毒的

opacity [əʊˈpæsɪtɪ] n. 不透明；不传导

opaque [əʊˈpeɪk] n. 不透明物 adj. 不透明的；不传热的；迟钝的 vt. 使不透明；使不反光

orbital [ˈɔːbɪtəl] adj. 轨道的

paramagnetism [ˌpærəˈmægnetɪzm] n. [物]顺磁性

permeability [ˌpɜːmɪəˈbɪlɪtɪ] n. 磁导率

perovskite [pəˈrɒvzkaɪt] n. [矿物]钙钛矿

phosphoresce [ˌfɒsfəˈres] vi. 发出磷光

phosphorous [ˈfɒsfərəs] adj. 磷的，含磷的；发磷光的

piezoelectricity [paɪˌiːzəʊɪlekˈtrɪsɪtɪ] n. 压电（现象）

polarization [ˌpəʊləraɪˈzeɪʃən] n. 极化；偏振；两极分化

porcelain [ˈpɔːsəlɪn] n. 瓷；瓷器 adj. 瓷制的

porosity [pɔːˈrɒsɪtɪ] n. 有孔性，多孔性

propagation [ˌprɒpəˈgeɪʃən] n. 传播；繁

殖;增殖

pyrex ['paɪəreks] n. 派热克斯玻璃(一种耐热玻璃)

quantum ['kwɒntəm] n. 量子论

radiation [ˌreɪdɪ'eɪʃən] n. 辐射,放射物,辐射

reflectivity [ˌriːflek'tɪvɪtɪ] n. [物]反射率;[光]反射性;反射比

refraction [rɪ'frækʃən] n. 折射;折光

remanence ['remənəns] n. [电磁]剩磁;剩余;剩余物

reradiation [ˌriːreɪdɪ'eɪʃən] n. [物]再辐射

samarium [sə'meərɪəm] n. [化学]钐

spin [spɪn] n. 旋转 vi. 旋转 vt. 使旋转

translucency [trænz'ljuːsənsɪ] n. 半透明

transmissivity [ˌtrænzmɪ'sɪvɪtɪ] n. [物]透射率;透光度;过滤系数

trivalent [traɪ'veɪlənt] adj. 三价的

ultraviolet [ˌʌltrə'vaɪəlet] adj. 紫外的;紫外线的 n. 紫外线辐射,紫外光

velocity [və'lɒsətɪ] n. [物]速度

yttrium ['ɪtrɪəm] n. [化学]钇

Exercises

1. Translate the following Chinese phrases into English

(1) 磁化率　　　(2) 外场　　　(3) 光学现象

(4) 绝缘材料　　(5) 固体材料　(6) 价电子

(7) 电场　　　　(8) 周期表　　(9) 热膨胀

(10) 热冲击　　(11) 内部温度分布　(12) 电子转换

2. Translate the following English phrases into Chinese

(1) relative permeability　　(2) magnetic field

(3) electrical conduction　　(4) energy band

(5) quantum theory　　　　(6) forbidden band

(7) n-type semiconductors　(8) heat capacity

(9) thermal conductivity　　(10) thermal stresses

(11) optical property

3. Translate the following Chinese sentences into English

(1) 近年来,高温超导体的应用已经超出了实验室的范围。

(2) 在可见光范围内,不同波长引起不同的颜色感觉。

(3) 超导性意味着零电阻,也就意味着没有能量的浪费。

(4) 硅是半导体工业的基础材料。

(5) 霍尔效应是一种发现、研究和应用都很早的磁电效应。

(6) 严格区别铁磁性与顺磁性物质有时是很困难的。

4. Translate the following English sentences into Chinese

(1) Magnetic nanocomposites with 5-10 fold increases in magnetocaloric effects should be used to develop magnetic refrigerators that operate at room temperature.

Chapter 9 Physical Properties of MATERIALS

(2) Donor: The impurities provide an energy level with electrons in band gaps. This kind of semiconductors is conductive mainly depending on electrons, which is Called n-type semiconductor.

(3) At high temperature, the semiconductor is not different from the conductor in the conductivity.

(4) At room temperature, all the metals are paramagnetic.

(5) Electron spin is essential in understanding many atomic phenomena.

5. Translate the following Chinese essay into English

随着粒子尺寸的减小,粒子中的磁畴数就会减少。一般地,多畴粒子的矫顽力小于单畴粒子的矫顽力。单畴粒子的矫顽力是由所谓的磁晶各向异性与形状各向异性共同决定的,因此,细长的单畴粒子优先应用于磁记录材料。具有很高的饱和磁化强度、适中矫顽力的单畴粒子适合于作为磁记录材料。但是,当粒子变得足够小,由于热扰动,磁矩不可能有择优取向,而展示出超顺磁性。

6. Translate the following English essay into Chinese

Piezoelectric ceramic fibers, given their unique properties of flexibility, light weight and higher output per pound of material, offer the greatest potential for enabling the wide-scale deployment of self-powered piezoelectric ceramic systems.

扫一扫,查看更多资料

Chapter 10 Chemical Properties of Materials

10.1 Introduction

To one degree or another, most materials experience some type of interaction with a large number of diverse environments. Often, such interactions impair a material's usefulness as a result of the deterioration of its mechanical properties (e. g., ductility and strength), other physical properties, or appearance.

Deteriorative mechanisms are different for the three material types. In metals, there is actual material loss either by dissolution (corrosion) or by the formation of nonmetallic scale or film (oxidation). Ceramic materials are relatively resistant to deterioration, which usually occurs at elevated temperatures or in rather extreme environments; the process is frequently also called corrosion. For polymers, mechanisms and consequences differ from those for metals and ceramics, and the term degradation is most frequently used. Polymers may dissolve when exposed to a liquid solvent, or they may absorb the solvent and swell; also electromagnetic radiation (primarily ultraviolet) and heat may cause alterations in their molecular structures.

10. 2 Corrosion of Metals

Corrosion is defined as the destructive and unintentional attack of a metal; it is electrochemical and ordinarily begins at the surface. The problem of metallic corrosion is one of significant proportions; in economic terms. It has been estimated that approximately 5 % of an industrialized nation's income is spent on corrosion prevention and the maintenance or replacement of products lost or contaminated as a result of corrosion reactions. The consequences of corrosion are all too common.

Chapter 10 Chemical Properties of Materials

Familiar examples include the rusting of automotive body panels and radiator and exhaust components.

10.2.1 Cost of Corrosion in Industry

The annual cost of corrosion in industry is surprising. Please see the Table 10.1.

Table 10.1 Corrosion expense statistic for some country in 1984 years

Country	Expense (hundred million)	Total produce value / %	Country	Expense (hundred million)	Total produce value / %
America	750 $	4	England	100 £	3.5
USSR	147 Rbl	2	France	1 150 FF	1.5
gfr	300 DM	3	Japan	133~150 $	1.3

It is common knowledge in our life that the rusting of steel is phenomenon which plagues not only the owner of the domestic "tin lizzie[①]" but costs us in Britain alone something in excess of £500 million annually in the application of protective measures.

10.2.2 Classification of Corrosion

Corrosion has been classified in many different ways. Common classification of corrosion has been as Table 10.2.

Table 10.2 Classification of corrosion

Corrosion process	Corrosion shape		Environmental corrosion	
	General corrosion	Localized corrosion	Dry corrosion	Wet corrosion
Chemical corrosion		Pitting		Atmospheric
Electrochemical corrosion		Crevice		Soil
Physical corrosion		Galvanic		Sea water
		Inter-granular		Microbial
		Stress		Acid, alkali and salt liquid
		Hydrogen embrittlement		
		Fatigue		
		Selective		

Chapter 10 Chemical Properties of Materials

Wet corrosion occurs when a liquid is present. This usually involves aqueous or electrolytes and accounts for the greatest amount of corrosion by far. A common example is corrosion of steel by water. Dry corrosion occurs in absence of a liquid phase or above the dew point of the environment. Vapors and gases are usually the corrosion. Dry corrosion is most often associated with high temperatures. An example is attack on steel by furnace gases.

10.2.3 Corrosion Mechanism

(1) Dry corrosion (or oxidation of metals)

Why the materials have happen corrosion? The dry corrosion mechanism may be explained that it is an oxidation process in atmosphere. For example, whilst many metals tend to oxidize to some extent at all temperatures, most engineering metals do not scale appreciably except at high temperatures. When iron is heated strongly in an atmosphere containing oxygen it becomes coated with a film of black scale. A chemical reaction has taken place between atoms of iron molecules of atmospheric oxygen:

$$2Fe + O_2 \longrightarrow 2FeO \qquad (10-1)$$

Although the reaction can be expressed by the above simple equation in fact atoms of iron have been oxidized whilst atoms of oxygen have been reduced. These processes are associated with a transfer of electrons from one atom to the other.

$$\text{Oxidation } Fe \longrightarrow Fe^{++} + 2 \text{ electrons }(e^-)$$
$$\text{Reduction } O + 2 \text{ electrons }(e^-) \longrightarrow O^{--} \qquad (10-2)$$

It should be noted that, in the chemical sense, the terms oxidation and reduction have a wider meaning than the combination or separation of a substance with oxygen. Thus, a substance is oxidized if its atoms lose electrons whilst it is reduced if its atoms (or groups of atoms) gain electrons. For this reason we say that iron is oxidized if it combines with sulphur, chlorine or any other substance which will accept electrons from the atoms of iron.

In some cases the formation of the oxide film protects the metal from further oxidation yet in many instances oxidation and scaling continue. For this to occur either molecules of oxygen must pass through a very porous film of oxide, or ions of either oxygen of the metal must migrate within a continuous film.

Figure 10.1 indicates the state of affairs which exists within the film of iron oxide scale. The positively charged Fe^{++} ions are attracted outwards towards the cathodic regions, $i.e.$ those rich in negative charge, whilst the negatively charged O^{--} ions are attracted inwards towards the region which is richer in positive charge, i. e. that near the metal surface. Since metallic ions are generally smaller than oxygen ions, the diffusion of the metal ions outwards is quicker than the diffusion of oxygen ions inwards. The rate of oxidation of the metal will depend partly on the mobility of the ions through the oxide film but also upon the rate of flow of electrons outwards. These mobilities are in turn affected by the nature and structure of the film, particularly in terms of "vacancies", that is, positions from which ions are missing. The number of vacancies is affected by the presence of solute atoms such as chromium and ionic mobilities are reduced as a result.

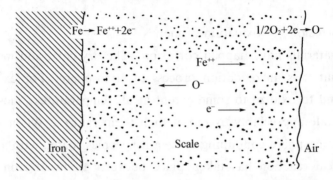

Figure 10.1 **The mechanism of the scaling of iron at high temperatures**

As mentioned above oxidation rates also depend upon the porosity of the oxide film. Some metals form ions which are much smaller than the original atoms. Consequently as ions are formed there is a volume reduction of the scale which, as a result, becomes porous and permits easier access of oxygen to the metal surface. Contraction of the oxide film may also cause it to flake off thus exposing a fresh metal surface. In some cases expansion of the film may occur causing it to buckle and become detached.

An oxide film which adheres tightly to the surface of the metal generally offers good protection. Good adhesion is the result of coherence between the film and the metal beneath. In Figure 10.2 (a) there is good "matching" between the ions in the metal surface and the metallic ions in the oxide film such that the structure is virtually continuous, whilst in Figure 10.2 (b) matching is absent so that the two surfaces will be non-coherent and there will be little adhesion. The high degree of coherence between aluminum and its oxide leads to the effective protection of this metal especially when the film is artificially thickened by anodizing.

Chapter 10　Chemical Properties of Materials

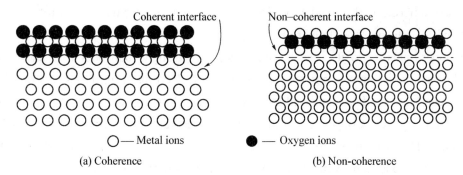

○— Metal ions　　●— Oxygen ions

(a) Coherence　　(b) Non-coherence

Figure 10.2 Adhesion between a metal surface and the covering oxide film. In the row of metal ions at the interface "matches up" with those in the oxide film (a), but in ions are "out of step" at the interface (b)

Some alloys are prone to attack by atmospheres containing sulphurous gases. Since nickel readily forms a sulphide at high temperatures it is particularly liable to be oxidized by gases containing sulphur. Heatresisting steels used in such conditions must be of the high-chromium type, preferably containing little or no nickel.

(2) Wet corrosion (or electrochemical oxidation)

What is wet corrosion? What's wet corrosion mechanism?

Iron dose not rust in a completely dry atmosphere nor will it rust in completely pure, oxygen-free water[②], but in a moist atmosphere the well-known reddish-brown deposit of ferric hydroxide soon begins to develop. The overall chemical reaction representing resting can be expressed by a simple chemical equation:

$$4Fe + 6H_2O + 3O_2 \longrightarrow 4Fe(OH)_3 \qquad (10-3)$$

This general result, however, is achieved in a number of stages and the fundamental principle involved is that atoms of iron in contact with oxygen and water are oxidized, that is they lose electrons and enter solution as ferrous ions (Fe^{++}):

$$Fe \longrightarrow Fe^{++} + 2 \text{ electrons } (2e^-) \qquad (10-4)$$

These ferrous ions are ultimately oxidized further to ferric ions (Fe^{+++}) by the removal of another electron:

$$Fe^{++} \longrightarrow Fe^{+++} + e^- \qquad (10-5)$$

As iron goes into solution in the form of ions the corresponding electrons are released. These electrons immediately combine with other ions so that overall equilibrium is maintained.

The ease with which a metal can be oxidized in this way depends upon the ease with which valency electrons can be removed from its atoms. Thus metals like calcium, aluminium and zinc hold their valency electrons comparatively loosely and can therefore be oxidized more easily than iron; but the noble metals gold, silver and platinum retain their valency electrons more strongly and are therefore much more difficult to oxidize than iron.

Copper holds on to its valency electrons more strongly than does iron and, under suitable conditions, copper ions will "steal" valency electrons from atoms of iron. Thus when a pen-knife blade is immersed in copper sulphate solution the blade becomes coated with metallic copper. Copper ions at the surface of the blade have removed electrons from the atoms of iron there so that the resultant ferrous ions have gone into solution thus replacing the copper ions which have been deposited as copper atoms.

$$Fe \longrightarrow Fe^{++} + 2e^- \tag{10-6}$$

$$Cu^{++} + 2e^- \longrightarrow Cu \tag{10-7}$$

Copper sulphate solution is sometimes used to coat the surface of steel with a thin layer of copper as an aid to "marking out[③]", in the tool room.

10.2.4 Electrochemical Considerations

For metallic materials, the corrosion process is normally electrochemical, that is, a chemical reaction in which there is transfer of electrons from one chemical species to another. Metal atoms characteristically lose or give up electrons in what is called an oxidation reaction. Examples in which metals oxidize are

$$Fe \longrightarrow Fe^{2+} + 2e^- \tag{10-8a}$$

$$Al \longrightarrow Al^{3+} + 3e^- \tag{10-8b}$$

The site, at which the oxidation takes place, is called the anode. The electrons generated from each metal atom that is oxidized must be transferred to and become a part of another chemical species in what is termed a reduction reaction. For example, some metals undergo corrosion in acid solutions, which have a high concentration of hydrogen (H^+) ions; the H^+ ions are reduced as hydrogen gas. Other reduction reactions are possible, depending on the nature of the solution to which the metal is exposed. For an acid solution having dissolved oxygen, reduction according to

$$O_2 + 4H^+ + 4e^- \longrightarrow 2H_2O \tag{10-9}$$

will probably occur. Or, for a neutral or basic aqueous solution in which oxygen is also dissolved,

$$O_2 + 2H_2O + 4e^- \longrightarrow 4OH^- \tag{10-10}$$

The location at which reduction occurs is called the cathode. Furthermore, it is possible for two or more of the reduction reactions above to occur simultaneously.

As a consequence of oxidation, the metal ions may either go into the corroding solution as ions, or they may form an insoluble compound with nonmetallic elements.

Not all metallic materials oxidize to form ions with the same degree of ease. Metallic materials may be rated as to their tendency to experience oxidation when coupled to other metals in solutions of their respective ions. Table 10.2 represents the corrosion tendencies for the several metals; those at the top (i. e., gold and platinum) are noble, or chemically inert. Moving down the table, the metals become increasingly more active, that is, more susceptible to oxidation. Sodium and potassium have the highest reactivities.

Most metals and alloys are subject to oxidation or corrosion to one degree or another in a wide variety of environments; that is, they are more stable in an ionic state than as metals. In thermodynamic terms, there is a net decrease in free energy in going from metallic to oxidized states. Consequently, essentially all metals occur in nature as compounds-for example, oxides, hydroxides, carbonates, silicates, sulfides and sulfates. Two notable exceptions are the noble metals gold and platinum. For them, oxidation in most environments is not favorable, and, therefore, they may exist in nature in the metallic state.

Even though Table 10.2 was generated under highly idealized conditions and has limited utility, it nevertheless indicates the relative reactivities of the metals. A more realistic and practical ranking, however, is provided by the galvanic series, Table 10.3. This represents the relative reactivities of a number of metals and commercial alloys in seawater. The alloys near the top are cathodic and unreactive, whereas those at the bottom are most anodic; no voltages are provided.

Table 10.2 The corrosion tendencies for common metals

	Electrode reaction
↑	$Au^{3+} + 3e^- \longrightarrow Au$
	$O_2 + 4H^+ + 4e^- \longrightarrow 2H_2O$
	$Pt^{2+} + 2e^- \longrightarrow Pt$
	$Ag^+ + e^- \longrightarrow Ag$
	$Fe^{3+} + e^- \longrightarrow Fe^{2+}$
Increasingly inter (cathodic)	$O_2 + 2H_2O + 4e^- \longrightarrow 4OH^-$
	$Cu^{2+} + 2e^- \longrightarrow Cu$
	$2H^+ + 2e^- \longrightarrow H_2$
Increasingly active (anodic)	$Pb^{2+} + 2e^- \longrightarrow Pb$
	$Sn^{2+} + 2e^- \longrightarrow Sn$
	$Ni^{2+} + 2e^- \longrightarrow Ni$
	$Co^{2+} + 2e^- \longrightarrow Co$
↓	$Cd^{2+} + 2e^- \longrightarrow Cd$

Chapter 10 Chemical Properties of Materials

(to be continued)

	Electrode reaction
Increasingly inter (cathodic) ↑ Increasingly active (anodic) ↓	$Fe^{2+} + 2e^- \longrightarrow Fe$ $Cr^{3+} + 3e^- \longrightarrow Cr$ $Zn^{2+} + 2e^- \longrightarrow Zn$ $Al^{3+} + 3e^- \longrightarrow Al$ $Mg^{2+} + 2e^- \longrightarrow Mg$ $Na^+ + e^- \longrightarrow Na$ $K^+ + e^- \longrightarrow K$

Table 10.3 The galvanic series

Increasingly inter (cathodic) ↑ Increasingly active (anodic) ↓	Platinum Gold Graphite Titanium Silver ⎡316 Stainless steel (passive) ⎣304 Stainless steel (passive) ⎡Inconel (80Ni-13Cr-7Fe) (passive) ⎣Nickel (passive) ⎡Monel (70Ni-30Cu) Copper-nickel alloys Bronzes (Cu-Sn alloys) Copper ⎣Brasses (Cu-Zn alloys) ⎡Inconel (active) ⎣Nickel (active) Tin Lead ⎡316 Stainless steel (active) ⎣304 Stainless steel (active) ⎡Cast iron ⎣Iron and steel Aluminum alloys Cadmium Commercially pure aluminum Zinc Magnesium and magnesium alloys

10.2.5 Corrosion Rates

The corrosion rate, or the rate of material removal as a consequence of the chemical action, is an important corrosion parameter. This may be expressed as the corrosion penetration rate (CPR), or the thickness loss of material per unit of time. The formula for this calculation is

$$\text{CPR} = \frac{KW}{\rho At} \qquad (10-11)$$

where W is the weight loss after exposure time t; ρ and A represent the density and exposed specimen area, respectively, and K is a constant, its magnitude depending on the system of units used. The CPR is conveniently expressed in terms of millimeters per year (mm/a).

10.2.6 Passivity

Some normally active metals and alloys, under particular environmental conditions, lose their chemical reactivity and become extremely inert. This phenomenon, termed passivity, is displayed by chromium, iron, nickel, titanium and many of their alloys. It is felt that this passive behavior results from the formation of a highly adherent and very thin oxide film on the metal surface, which serves as a protective barrier to further corrosion. Stainless a result of steels are highly resistant to corrosion in a rather wide variety of atmospheres as a passivation. They contain at least 11% chromium that, as a solid-solution alloying element in iron, minimizes the formation of rust; instead, a protective surface film forms in oxidizing atmospheres. (Stainless steels are susceptible to corrosion in some environments, and therefore are not always "stainless".) Aluminum is highly corrosion resistant in many environments because it also passivates. If damaged, the protective film normally reforms very rapidly. However, a change in the character of the environment (e. g., alteration in the concentration of the active corrosive species) may cause a passivated material to revert to an active state. Subsequent damage to a preexisting passive film could result in a substantial increase in corrosion rate, by as much as 100 000 times.

10.2.7 Environmental Effects

The variables in the corrosion environment, which include fluid velocity, temperature and composition, can have a decided influence on the corrosion properties of the materials that are in contact with it. In most instances, increasing fluid velocity enhances the rate of corrosion due to erosive effects. The rates of most chemical reactions rise with increasing temperature; this also holds for the great majority of corrosion situations. Increasing the concentration of the corrosive species (e. g., H^+ ions in acids) in many situations produces a more rapid rate of corrosion. However, for materials capable of passivation, raising the corrosive content may result in an active-to-passive transition, with a considerable reduction in corrosion.

Cold working or plastically deforming ductile metals is used to increase their strength; however, a cold-worked metal is more susceptible to corrosion than the same material in an annealed state. For example, deformation processes are used to shape the head and point of a nail; consequently, these positions are anodic with respect to the shank region.

10.2.8 Forms of Corrosion

It is convenient to classify corrosion according to the manner in which it is manifest. Metallic corrosion is sometimes classified into eight forms: uniform, galvanic, crevice, pitting, intergranular, selective leaching, erosion-corrosion and stress corrosion. In addition, hydrogen embrittlement is, in a strict sense, a type of failure rather than a form of corrosion; however, it is often produced by hydrogen that is generated from corrosion reactions.

(1) Uniform attack

Uniform attack is a form of electrochemical corrosion that occurs with equivalent intensity over the entire exposed surface and often leaves behind a scale or deposit. In a microscopic sense, the oxidation and reduction reactions occur randomly over the surface. Some familiar examples include general rusting of steel and iron and the tarnishing of silverware. This is probably the most common form of corrosion. It is also the least objectionable because it can be predicted and designed for with relative ease.

(2) Galvanic corrosion

Galvanic corrosion occurs when two metals or alloys having different compositions are electrically coupled while exposed to an electrolyte. The less noble or more reactive metal in the particular environment will experience corrosion; the more inert metal, the cathode, will be protected from corrosion. For example, steel screws corrode when in contact with brass in a marine environment; or if copper and steel tubing are joined in a domestic water heater, the steel will corrode in the vicinity of the junction. Depending on the nature of the solution, one or more of reduction reactions will occur at the surface of the cathode material.

The rate of galvanic attack depends on the relative anode-to-cathode surface areas that are exposed to the electrolyte, and the rate is related directly to the cathode-anode area ratio; that is, for a given cathode area, a smaller anode will corrode more rapidly than a larger one.

A number of measures may be taken to significantly reduce the effects of galvanic corrosion. These include the following:

1) If coupling of dissimilar metals is necessary, choose two that are close together in the galvanic series.

2) Avoid an unfavorable anode-to-cathode surface area ratio; use an anode area as large as possible.

3) Electrically insulate dissimilar metals from each other.

4) Electrically connect a third, anodic metal to the other two; this is a form of cathodic protection.

(3) Crevice corrosion

Electrochemical corrosion may also occur as a consequence of concentration differences of ions or dissolved gases in the electrolyte solution, and between two regions of the same metal piece. For such a concentration cell, corrosion occurs in the locale that has the lower concentration. A good example of this type of corrosion occurs in crevices and recesses or under deposits of dirt or corrosion products where the solution becomes stagnant and there is localized depletion of dissolved oxygen. Corrosion preferentially occurring at these positions is called crevice corrosion (Figure 10.3). The crevice must be wide enough for the solution to penetrate, yet narrow enough for stagnancy; usually the width is several thousandths of an inch.

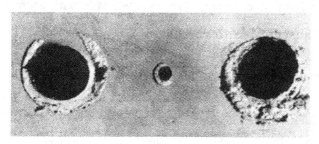

Figure 10.3 On this plate, which was immersed in seawater, crevice corrosion has occurred at the regions that were covered by washers

Chapter 10 Chemical Properties of Materials

Crevice corrosion may be prevented by using welded instead of riveted or bolted joints, using nonabsorbing gaskets when possible, removing accumulated deposits frequently, and designing containment vessels to avoid stagnant areas and ensure complete drainage.

(4) Pitting

Pitting is another form of very localized corrosion attack in which small pits or holes form. They ordinarily penetrate from the top of a horizontal surface downward in a nearly vertical direction. It is an extremely insidious type of corrosion, often going undetected and with very little material loss until failure occurs. An example of pitting corrosion is shown in Figure 10.4.

Figure 10.4 The pitting of a 304 stainless steel plate by an acid-chloride solution

The mechanism for pitting is probably the same as for crevice corrosion in that oxidation occurs within the pit itself, with complementary reduction at the surface. It is supposed that gravity causes the pits to grow downward, the solution at the pit tip becoming more concentrated and dense as pit growth progresses. A pit may be initiated by a localized surface defect such as a scratch or a slight variation in composition. In fact, it has been observed that specimens having polished surfaces display a greater resistance to pitting corrosion. Stainless steels are somewhat susceptible to this form of corrosion; however, alloying with about 2 % molybdenum enhances their resistance significantly.

(5) Intergranular corrosion

As the name suggests, intergranular corrosion occurs preferentially along grain boundaries for some alloys and in specific environments. The net result is that a macroscopic specimen disintegrates along its grain boundaries. This type of corrosion is especially prevalent in some stainless steels. When heated to temperatures between 500 and 800 °C for sufficiently long time periods, these alloys become sensitized to intergranular attack. It is believed that this heat treatment permits the formation of small precipitate particles of chromium carbide ($Cr_{23}C_6$)

by reaction between the chromium and carbon in the stainless steel. These particles form along the grain boundaries, as illustrated in Figure 10.5. Both the chromium and the carbon must diffuse to the grain boundaries to form the precipitates, which leaves a chromium-depleted zone① adjacent to the grain boundary. Consequently, this grain boundary region is now highly susceptible to corrosion.

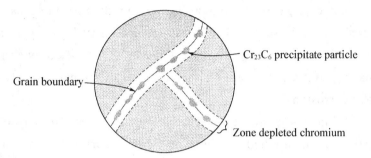

Figure 10.5 Schematic illustration of chromium carbide particles that have precipitated along grain boundaries in stainless steel, and the attendant zones of chromium depletion

Intergranular corrosion is an especially severe problem in the welding of stainless steels, when it is often termed weld decay.

Stainless steels may be protected from intergranular corrosion by the following measures:

1) Subjecting the sensitized material to a high-temperature heat treatment in which all the chromium carbide particles are redissolved.

2) Lowering the carbon content below 0.03 wt % C so that carbide formation is minimal.

3) Alloying the stainless steel with another metal such as niobium or titanium, which has a greater tendency to form carbides than does chromium so that the Cr remains in solid solution.

(6) Selective leaching

Selective leaching is found in solid solution alloys and occurs when one element or constituent is preferentially removed as a consequence of corrosion processes. The most common example is the dezincification of brass, in which zinc is selectively leached from a copper-zinc brass alloy. The mechanical properties of the alloy are significantly impaired, since only a porous mass of copper remains in the region that has been dezincified. In addition, the material changes from yellow to a red or copper color. Selective leaching may also occur with other alloy systems in which aluminum, iron, cobalt, chromium and other elements are vulnerable to preferential removal.

(7) Erosion-corrosion

Erosion-corrosion arises from the combined action of chemical attack and mechanical abrasion or wear as a consequence of fluid motion. Virtually all metal alloys, to one degree or another, are susceptible to erosion-corrosion. It is especially harmful to alloys that passivate by forming a protective surface film; the abrasive action may erode away the film, leaving exposed a bare metal surface. If the coating is not capable of continuously and rapidly reforming as a protective barrier, corrosion may be severe. Relatively soft metals such as copper and lead are also sensitive to this form of attack. Usually it can be identified by surface grooves and waves having contours that are characteristic of the flow of the fluid.

(8) Stress corrosion

Stress corrosion, sometimes termed stress corrosion cracking, results from the combined action of an applied tensile stress and a corrosive environment; both influences are necessary. In fact, some materials that are virtually inert in a particular corrosive medium become susceptible to this form of corrosion when a stress is applied. Small cracks form and then propagate in a direction perpendicular to the stress (Figure 10.6), with the result that failure may eventually occur. Failure behavior is characteristic of that for a brittle material, even though the metal alloy is intrinsically ductile. Furthermore, cracks may form at relatively low stress levels, significantly below the tensile strength. Most alloys are susceptible to stress corrosion in specific environments, especially at moderate stress levels. For example, most stressed stainless steels corrode in solutions containing chloride ions, whereas brasses are especially vulnerable when exposed to ammonia.

Figure 10.6 A bar of steel that has been bent into a "horseshoe" shape using a nut-and-bolt assembly. While immersed in seawater, stress corrosion cracks formed along the bend at those regions where the tensile stresses are the greatest

The stress that produces stress corrosion cracking need not be externally applied; it may be a residual one that results from rapid temperature changes and uneven contraction, or for two-phase alloys in which each phase has a different coefficient of expansion. Also, gaseous and solid corrosion products that are entrapped internally can give rise to internal stresses.

Probably the best measure to take in reducing or totally eliminating stress corrosion is to lower the magnitude of the stress. This may be accomplished by reducing the external load or increasing the cross sectional area perpendicular to the applied stress. Furthermore, an appropriate heat treatment may be used to anneal out any residual thermal stresses.

(9) Hydrogen embrittlement

Various metal alloys, specifically some steels, experience a significant reduction in ductility and tensile strength when atomic hydrogen (H) penetrates into the material. This phenomenon is aptly referred to as hydrogen embrittlement. Strictly speaking, hydrogen embrittlement is a type of failure; in response to applied or residual tensile stresses, brittle fracture occurs catastrophically as cracks grow and rapidly propagate. Hydrogen in its atomic form (H as opposed to the molecular form, H_2) diffuses interstitially through the crystal lattice, and concentrations as low as several parts per million can lead to cracking. Furthermore, hydrogen induced cracks are most often transgranular, although intergranular fracture is observed for some alloy systems. A number of mechanisms have been proposed to explain hydrogen embrittlement; most of them are based on the interference of dislocation motion by the dissolved hydrogen.

For hydrogen embrittlement to occur, some source of hydrogen must be present, and, in addition, the possibility for the formation of its atomic species. Situations wherein these conditions are met include the following: pickling of steels (Pickling is a procedure used to remove surface oxide scale from steel pieces by dipping them in a vat of hot, dilute sulfuric or hydrochloric acid in sulfuric acid); electroplating; and the presence of hydrogen-bearing atmospheres (including water vapor) at elevated temperatures such as during welding and heat treatments. Also, the presence of what are termed "poisons" such as sulfur (i. e., H_2S) and arsenic compounds accelerates hydrogen embrittlement; these substances retard the formation of molecular hydrogen and thereby increase the residence time of atomic hydrogen on the metal surface. Hydrogen sulfide, probably the most aggressive poison, is found in petroleum fluids, natural gas. oil-well brines and geothermal fluids.

Some of the techniques commonly used to reduce the likelihood of hydrogen embrittlement include reducing the tensile strength of the alloy via a heat treatment, removal of the source of hydrogen, "baking" the alloy at an elevated temperature to drive out any dissolved hydrogen, and substitution of a more embrittlement resistant alloy.

10.2.9 Corrosion Environments

Corrosive environments include the atmosphere, aqueous solutions, soils, acids, bases, inorganic solvents, molten salts, liquid metals, and, last but not least, the human body. On a tonnage basis, atmospheric corrosion accounts for the greatest losses. Moisture containing dissolved oxygen is the primary corrosive agent, but other substances, including sulfur compounds and sodium chloride, may also contribute. This is especially true of marine atmospheres, which are highly corrosive because of the presence of sodium chloride. Dilute sulfuric acid solutions (acid rain) in industrial environments can also cause corrosion problems. Metals commonly used for atmospheric applications include alloys of aluminum and copper, and galvanized steel.

Water environments can also have a variety of compositions and corrosion characteristics. Freshwater normally contains dissolved oxygen, as well as other minerals several of which account for hardness. Seawater contains approximately 3.5% salt (predominantly sodium chloride), as well as some minerals and organic matter. Seawater is generally more corrosive than freshwater, frequently producing pitting and crevice corrosion. Cast iron, steel, aluminum, copper, brass and some stainless steels are generally suitable for freshwater use, whereas titanium, brass, some bronzes, copper-nickel alloys and nickel-chromium-molybdenum alloys are highly corrosion re resistant in seawater.

Soils have a wide range of compositions and susceptibilities to corrosion. Compositional variables include moisture, oxygen, salt content, alkalinity and acidity, as well as the presence of various forms of bacteria. Cast iron and plain carbon steels, both with and without protective surface coatings, are found most economical for underground structures.

10.2.10 Corrosion Prevention

Some corrosion prevention methods were treated relative to the eight forms of corrosion; however, only the measures specific to each of the various corrosion types were discussed.

Physical barriers to corrosion, as a general technique, are applied on surfaces in the form of films and coatings. A large diversity of metallic and nonmetallic coating materials is available. It is essential that the coating maintain a high degree of surface adhesion, which undoubtedly requires some preapplication surface treatment. In most cases, the coating must be virtually nonreactive in the corrosive environment and resistant to mechanical damage that exposes the bare metal to the corrosive environment. All three material types (metals, ceramics and polymers) are used as coatings for metals.

Another effective method of corrosion prevention is cathodic protection; it can be used for all eight different forms of corrosion as discussed above, and may, in some situations, completely stop corrosion. One cathodic protection technique employs a galvanic couple: the metal to be protected is electrically connected to another metal that is more reactive in the particular environment. The latter experiences oxidation, and, upon giving up electrons, protects the first metal from corrosion. The oxidized metal is often called a sacrificial anode, and magnesium and zinc are commonly used as such because they lie at the anodic end of the galvanic series. The process of galvanizing is simply one in which a layer of zinc is applied to the surface of steel by hot dipping. In the atmosphere and most aqueous environments, zinc is anodic to and will thus cathodically protect the steel if there is any surface damage. Any corrosion of the zinc coating will proceed at an extremely slow rate because the ratio of the anode-to-cathode surface area is quite large.

10.3 Corrosion of Ceramic Materials

Ceramic materials, being compounds between metallic and nonmetallic elements, may be thought of as having already been corroded. Thus, they are exceedingly immune to corrosion by almost all environments, especially at room temperature. Corrosion of ceramic materials generally involves simple chemical dissolution, in contrast to the electrochemical processes found in metals, as described above.

Ceramic materials are frequently utilized because of their resistance to corrosion. Glass is often used to contain liquids for this reason. Refractory ceramics must not only withstand high temperatures and provide thermal insulation but, in many instances, must also resist high-temperature attack by molten metals, salts, slags and glasses. Some of the new technology schemes for converting energy from one form to another that is more useful require relatively high temperatures, corrosive atmospheres and pressures above the ambient. Ceramic materials are much better suited to withstand most of these environments for reasonable time periods than are metals.

10.4 Degradation of Polymers

Polymeric materials also experience deterioration by means of environmental interactions. However, an undesirable interaction is specified as degradation rather than corrosion because the processes are basically dissimilar. Whereas most metallic corrosion reactions are electrochemical, by contrast, polymeric degradation is physiochemical; that is, it involves physical as well as chemical phenomena. Furthermore, a wide variety of reactions and adverse consequences are possible for polymer degradation. Polymers may deteriorate by swelling and dissolution. Covalent bond rupture, as a result of heat energy, chemical reactions and radiation is also possible, ordinarily with an attendant reduction in mechanical integrity. It should also be mentioned that because of the chemical complexity of polymers, their degradation mechanisms are not well understood.

10.4.1 Swelling and Dissolution

When polymers are exposed to liquids, the main forms of degradation are swelling and dissolution. With swelling, the liquid or solute diffuses into and is absorbed within the polymer; the small solute molecules fit into and occupy positions among the polymer molecules. Thus the macromolecules are forced apart such that the specimen expands or swells. Furthermore, this increase in chain separation results in a reduction of the secondary intermolecular bonding forces; as a consequence, the material becomes softer and more ductile. The liquid solute also lowers the glass transition temperature and, if depressed below the ambient temperature, will cause a once strong material to become rubbery and weak.

Swelling may be considered to be a partial dissolution process in which there is only limited solubility of the polymer in the solvent. Dissolution, which occurs when the polymer is completely soluble, may be thought of as just a continuation of swelling. For example, many hydrocarbon rubbers readily absorb hydrocarbon liquids such as gasoline.

In general, increasing molecular weight, increasing degree of crosslinking and crystallinity, and decreasing temperature result in a reduction of these deteriorative processes.

10.4.2 Bond Rupture

Polymers may also experience degradation by a process termed scission-the severance or rupture of molecular chain bonds. This causes a separation of chain segments at the point of scission and a reduction in the molecular weight. Bond rupture may result from exposure to radiation or to heat, and from chemical reaction.

Certain types of radiation, like electron beams, X-rays, β- and γ-rays and ultraviolet (UV) radiation, possess sufficient energy to penetrate a polymer specimen and interact with the constituent atoms or their electrons. One such reaction is ionization, in which the radiation removes an orbital electron from a specific atom, converting that atom into a positively charged ion. As a consequence, one of the covalent bonds associated with the specific atom is broken, and there is a rearrangement of atoms or groups of atoms at that point. This bond breaking leads to either scission or crosslinking at the ionization site, depending on the chemical structure of the polymer and also on the dose of radiation. In day-to-day use, the greatest radiation damage to polymers is caused by UV irradiation. After prolonged exposure, most polymer films become brittle, discolor, crack and fail. For example, camping tents begin to tear, dashboards develop cracks, and plastic windows become cloudy.

Oxygen, ozone and other substances can cause or accelerate chain scission as a result of chemical reaction. This effect is especially prevalent in vulcanized rubbers that have doubly bonded carbon atoms along the backbone molecular chains, and that are exposed to ozone (O_3), an atmospheric pollutant. Ordinarily, if the rubber is in an unstressed state, a film will form on the surface, protecting the bulk material from any further reaction. However, when these materials are subjected to tensile stresses, cracks and crevices form and grow in a direction perpendicular to the stress; eventually, rupture of the material may occur. This is why the sidewalls on rubber bicycle tires develop cracks as they age. Apparently these cracks result from large numbers of ozone-induced scissions. Chemical degradation is a particular problem for polymers used in areas with high levels of air pollutants such as smog and ozone.

Thermal degradation corresponds to the scission of molecular chains at elevated temperatures; as a consequence, some polymers undergo chemical reactions in which gaseous species are produced. These reactions are evidenced by a weight loss of material; a polymer's thermal stability is a measure of its resilience to this

Chapter 10 Chemical Properties of Materials

decomposition. Thermal stability is related primarily to the magnitude of the bonding energies between the various atomic constituents of the polymer: higher bonding energies result in more thermally stable materials.

10.4.3 Weathering

Many polymeric materials serve in applications that require exposure to outdoor conditions. Any resultant degradation is termed weathering, which may, in fact, be a combination of several different processes. Under these conditions deterioration is primarily a result of oxidation, which is initiated by ultraviolet radiation from the sun. Some polymers such as nylon and cellulose are also susceptible to water absorption, which produces a reduction in their hardness and stiffness. Resistance to weathering among the various polymers is quite diverse. The fluorocarbons are virtually inert under these conditions; but some materials, including poly (vinyl chloride) and polystyrene, are susceptible to weathering.

◆ Notes

① tin lizzie 轻快小汽车
② oxygen-free water 无氧水
③ marking out 划线；标出
④ chromium-depleted zone 贫铬区

◆ Vocabulary

acidity [ə'sɪdɪtɪ] n. 酸度；酸性
alkalinity [ˌælkə'lɪnɪtɪ] n. [化学]碱度；[化学]碱性
anode ['ænəʊd] n. 阳极（电解）；正极（原电池）
bacteria [bæk'tɪərɪə] n. [微]细菌
cathode ['kæθəʊd] n. 阴极（电解）；负极（原电池）
chlorine ['klɔːriːn] n. [化学]氯
crevice ['krevɪs] n. 裂缝；裂隙
depletion [dɪ'pliːʃn] n. 消耗；损耗
deterioration [dɪˌtɪərɪə'reɪʃn] n. 恶化；退化；堕落
deteriorative [dɪ'tɪərɪəreɪtɪv] adj. 恶化的；颓废的；退化的；变质的

dezincification [diːˌzɪŋkəfə'keɪʃn] n. 脱锌
dissolution [ˌdɪsə'luːʃn] n. 分解，溶解
drainage ['dreɪnɪdʒ] n. 排水；排水系统；污水；排水面积
electrolyte [i'lektrəlaɪt] n. 电解液，电解质；电解
erosion [i'rəʊʒn] n. 侵蚀，腐蚀
erosive [i'rəʊsɪv] adj. 腐蚀的；冲蚀的；侵蚀性的
fluorocarbon [ˌflʊərə'kɑːbən] n. [有化]碳氟化合物
galvanic [gæl'vænɪk] adj. 电流的；使人震惊的；触电似的
gasket ['gæskɪt] n. [机]垫圈；[机]衬垫

Chapter 10 Chemical Properties of Materials

gasoline ['gæsəli:n] n. 汽油

heatresisting ['hi:trɪz'ɪstɪŋ] n. 耐热

hydrocarbon [,haɪdrəʊ'kɑ:bən] n. [有化]碳氢化合物

insidious [ɪn'sɪdɪəs] adj. 阴险的；隐伏的；暗中为害的；狡猾的

intergranular [,ɪntə'grænjələ] adj. 颗粒间的，晶粒间的

moisture ['mɒɪstfə] n. 水分；湿度；潮湿；降雨量

nonreactive [,nɒnrɪ'æktɪv] adj. [化学]不反应的；无电抗的

ozone ['əʊzəʊn] n. [化学]臭氧

passivate ['pæsɪveɪt] vt. 使钝化

passivation [,pæsɪ'veɪʃən] n. [化学]钝化；钝化处理

passivity [pæ'sɪvɪtɪ] n. 钝化，钝态

penetrate ['penɪtreɪt] vi. 渗透；刺入；看透 vt. 渗透；穿透；洞察

penetration [,penɪ'treɪʃən] n. 渗透；突破；侵入；洞察力

pitting ['pɪtɪŋ] n. 凹陷；金属表面的腐蚀；点状腐蚀 adj. 点状的

pollutant [pə'lju:tənt] n. 污染物

poly ['pɒlɪ] adj.（polyester 的缩写）聚酯的，涤纶的

polystyrene [,pɒlɪ'staɪri:n] n. [高分子]聚苯乙烯

potassium [pə'tæsɪəm] n. [化学]钾

rupture ['rʌptʃə] n. 破裂；决裂；疝气 vt. 使破裂；断绝；发生疝 vi. 破裂；发疝气

scission ['sɪʃən] n. 切断，分离；断开

slag [slæg] n. 炉渣；矿渣；熔渣 vt. 使成渣；使变成熔渣 vi. 变熔渣

smog [smɒg] n. 烟雾

stagnancy ['stægnənsɪ] n. 停滞；迟钝；萧条；不景气

stagnant ['stægnənt] adj. 停滞的；不景气的；污浊的；迟钝的

sulphate ['sʌlfeɪt] n. 硫酸盐 vt. 使与硫酸化合 vi. 硫酸盐化

sulphide ['sʌlfaɪd] n. 硫化物

sulphur ['sʌlfə] n. 硫黄；硫黄色 vt. 使硫化；用硫黄处理

sulphurous ['sʌlfərəs] adj. 含硫黄的

swelling ['swelɪŋ] n. 肿胀；膨胀；增大；涨水 v. 肿胀；膨胀；增多 adj. 膨胀的；肿大的；突起的

tarnishing ['tɑ:nɪʃɪŋ] n. [化工]锈蚀，[化工]锈污 v. 使生锈，沾污

tonnage ['tʌnɪdʒ] n. 吨位，载重量；船舶总吨数，排水量

transgranular [træns'grænjʊlə] adj. 穿晶的；晶内

unintentional [,ʌnɪn'tenʃənəl] adj. 非故意的；无意识的

velocity [və'lɒsətɪ] n. [物]速度

voltage ['vəʊltɪdʒ] n. [电]电压

weathering ['weðərɪŋ] n. [地质]风化作用 v. 风干；使褪色；经受住

◆ Exercises

1. Translate the following Chinese phrases into English

(1) 化学性能　　(2) 金属的氧化　　(3) 酸性溶液

(4) 均匀腐蚀　　(5) 缝隙腐蚀　　(6) 晶间腐蚀

Chapter 10 Chemical Properties of Materials

(7) 选择浸出   (8) 阴极保护   (9) 设备设计

(10) 制备方法

2. Translate the following English phrases into Chinese

(1) corrosion process   (2) ferrous ion

(3) environmental effect   (4) galvanic corrosion

(5) dissolved oxygen   (6) stainless steel

(7) hydrogen embrittlement   (8) bond rupture

(9) surface finish

3. Translate the following Chinese sentences into English

(1) 如果溶液中有溶解氧,会与电子反应生成氢氧根离子。反过来,氢氧根离子又会和金属离子反应。比如,生成常见的铁锈。

(2) 高硅铸铁具有很强的抗高温氧化能力。

(3) 维护费用中主要部分是由腐蚀引起的,腐蚀防护在很多设计中也是至关重要的。

(4) 新磨损的界面暴露在空气中时,会迅速生成厚约10埃的保护膜,这种保护膜可以有效地防止金属被腐蚀。

(5) 腐蚀的基本原理几乎可以涵盖所有情况下的腐蚀反应,但不包括高温下发生的某些腐蚀反应。

4. Translate the following English sentences into Chinese

(1) The key to the behavior of the metal is whether the oxide that does form creates a continuous, nonporous protective layer that prevents further oxidation or creates other corrosive attack on the metal.

(2) The inner surface of this kind of workpiece castings is often protected against corrosion with a special coat.

(3) Two different metals in contact with an electrolyte, typically water, compose an electrolytic cell, or galvanic cell.

(4) Since ceramics are often oxides, nitrides, or sulfides, compounds found in nature, there is little driving force for corrosion.

(5) The corrosion process is usually electrochemical in nature, having the essential features of a battery.

(6) The ions leave their corresponding negative charge in the form of electrons in the metal which travel to the location of the cathode through a conductive path.

5. Translate the following Chinese essay into English

自由基要得到另外一个电子以形成电子对的倾向使得它高度活泼。因此,它通过夺得一个电子而破坏另一个分子的价键,使它带有电子(另一个自由基)。

6. Translate the following English essay into Chinese

The bonding forms of crystals include electrostatic bonding, covalent binding, metal-binding and Van der Waals bonding. The bond forms of crystals have close relationships with their crystal structures, physics and chemical properties. The study on crystalline cohesion of crystals is thus the base to investigate material properties.

扫一扫,查看更多资料

Main References

[1] 肖建中. 材料科学导论[M]. 北京:中国电力出版社,2001.
[2] Jean P. Mercier, Gerald Zambelli, Wifried Kurz. *Introduction to Materials Science*[M]. Elsevier Publishing, 2002.
[3] V. S. R. Murthy, A. K. Jena, K. P. Gupta, et al. *Structure and Properties of Engineering Materials*[M]. Tata McGraw-Hill Publishing Company Limited, Kanpur, India, 2003.
[4] 张 军. 材料专业英语译写教程[M]. 北京:机械工业出版社,2005.
[5] 侯纯明. 物理化学双语基础[M]. 北京:中国石化出版社,2006.
[6] William D. Callister, Jr. Fundamentals of Materials Science and Engineering (Fifth Edition) [M]. 北京:化学工业出版社,2006.
[7] 黄培彦. 材料科学与工程导论(双语)[M]. 广州:华南理工大学出版社,2007.
[8] 范积伟. 材料专业英语[M]. 北京:机械工业出版社,2010.
[9] 陈克正,王 玮,刘春廷. 材料科学与工程导论(双语)[M]. 北京:化学工业出版社,2011.
[10] 水中和. 材料概论(双语)[M]. 武汉:武汉理工大学出版社,2012.
[11] 匡少平,王世颖. 材料科学与工程专业英语[M]. 北京:化学工业出版社,2013.
[12] 张耀君. 纳米材料基础(双语版)[M]. 北京:化学工业出版社,2013.
[13] 刘东平. 材料物理(双语)[M]. 大连:大连理工大学出版社,2014.
[14] 刘爱国. 材料科学与工程专业英语[M]. 哈尔滨:哈尔滨工业大学出版社,2015.
[15] 赵 杰. 材料科学基础(双语版)[M]. 大连:大连理工大学出版社,2015.